PUBLISHER COMMENTARY

We print NASA's handbooks and standards for the convenience of those that use them on a daily basis. We print all of these a full 8 ½ by 11 with large text so they are easy to read. Yes, color books are expensive to print so unless the information relies on the use of color for proper interpretation or understanding, we print most books in black and white to keep the cost down. All these documents are available for download for free from NASA, however printing them all over a network printer would take days.

Why buy a book you can download free? We print this so you don't have to.

All these books are available for free download from the government web site. Some are available only in electronic media. Some online docs are missing pages or barely legible.

We at 4th Watch Publishing are former government employees, so we know how government employees actually use the standards. When a new standard is released, an engineer prints it out, punches holes and puts it in a 3-ring binder. While this is not a big deal for a 5 or 10-page document, many NIST documents are over 100 pages and printing a large document is a time-consuming effort. So, an engineer that's paid $75 an hour is spending hours simply printing out the tools needed to do the job. That's time that could be better spent doing engineering. We publish these documents so engineers can focus on what they were hired to do – engineering. It's much more cost-effective to just order the latest version from Amazon.com

If there is a standard you would like published, let us know. Our web site is www.usgovpub.com

www.usgovpub.com

Copyright © 2019 4th Watch Publishing Co. All Rights Reserved

List of Other NASA Publications Available on Amazon.com:

NASA-STD-5001B	Structural Design and Test Factors of Safety for Spaceflight Hardware
NASA-STD-5006A	General Welding Requirements for Aerospace Materials
NASA-STD-5008B	Protective Coating of Carbon Steel, Stainless Steel, and Aluminum on Launch Structures, Facilities, and Ground Support Equipment
NASA-STD-5009A	Nondestructive Evaluation Requirements for Fracture-Critical Metallic Components
NASA-STD-5012B	Strength and Life Assessment Requirements for Liquid-Fueled Space Propulsion System Engines
NASA-STD-5019A	Fracture Control Requirements for Spaceflight Hardware
NASA-STD-5005D	Standard for The Design and Fabrication of Ground Support Equipment
NASA-HDBK-8739.21	Workmanship Manual for Electrostatic Discharge Control
NASA-HDBK 8739.23A	NASA Complex Electronics Handbook for Assurance Professionals (Color)
NASA-HDBK-8719.14	Handbook for Limiting Orbital Debris (Color)
NASA-HDBK-8709.22	Safety and Mission Assurance Acronyms, Abbreviations, and Definitions
NASA-HDBK-7009	NASA Handbook for Models and Simulations: An Implementation Guide For NASA-STD-7009 (Color)
NASA-HDBK-8739.19-2	Measuring and Test Equipment Specifications NASA Measurement Quality Assurance Handbook – Annex 2
NASA-HDBK-8739.19-3	Measurement Uncertainty Analysis Principles and Methods NASA Measurement Quality Assurance Handbook – Annex 3
NASA-HDBK-8739.19-4	Estimation and Evaluation of Measurement Decision Risk NASA Measurement Quality Assurance Handbook – Annex 4
NASA RCM	Reliability-Centered Maintenance Guide for Facilities and Collateral Equipment

www.usgovpub.com

NASA TECHNICAL STANDARD	NASA-STD-5005D
National Aeronautics and Space Administration Washington, DC 20546-0001	Approved: 06-14-2013 Superseding NASA-STD-5005C

STANDARD FOR THE DESIGN AND FABRICATION OF

GROUND SUPPORT EQUIPMENT

MEASUREMENT SYSTEM IDENTIFICATION:
METRIC/SI (ENGLISH)

DOCUMENT HISTORY LOG

Status	Document Revision	Approval Date	Description
Baseline		5-10-1996	Baseline Release
Revision	A	5-15-2001	Incorporated numerous updates to applicable and reference documents
Revision	B	9-15-2003	Incorporated numerous updates to applicable and reference documents
Interim	C	7-20-2007	Interim Revision (General Revision—Document Completely Rewritten) Incorporated numerous updates to applicable and reference documents, added M&P Section 5.11, restructured to single verifiable requirement per numbered paragraph, and moved non-verifiable best practices to italicized guidance statements.
Revision	C	3-13-2009	General Revision—Document Completely Rewritten Transitioned Interim NASA Technical Standard NASA-STD-(I)-5005C to NASA Technical Standard NASA-STD-5005C.
Revision	D	6-14-2013	General Revision—Document Completely Rewritten Incorporated numerous updates to applicable and reference documents, reduced the number of Center-specific standards in lieu of industry standards, and moved M&P requirements from 5.11 to Section 6.

FOREWORD

This Standard is published by the National Aeronautics and Space Administration (NASA) to provide uniform engineering and technical requirements for processes, procedures, practices, and methods that have been endorsed as standard for NASA programs and projects, including requirements for selection, application, and design criteria of an item.

This Standard is approved for use by NASA Headquarters and NASA Centers, including Component Facilities and Technical and Service Support Centers.

This Standard establishes minimum requirements and engineering best practices for design and development of ground systems and equipment intended for use in preparing space flight systems for flight after acceptance by the Government. This Standard does not apply to facilities or equipment used in the manufacturing of space flight systems.

Individual provisions of this Standard should be and are intended to be tailored (i.e., modified or deleted) by contract or program specifications to meet specific program, project, and Center needs and constraints based on a risk assessment by Safety and Mission Assurance (S&MA) according to program and Center procedures.

Requests for information, corrections, or additions to this Standard should be submitted via "Feedback" in the NASA Standards and Technical Assistance Resource Tool at https://standards.nasa.gov.

Origial Signed By: *06-14-2013*
_____   _____

Michael G. Ryschkewitsch   Approval Date
NASA Chief Engineer

TABLE OF CONTENTS

TABLE OF CONTENTS (Continued)

TABLE OF CONTENTS (Continued)

TABLE OF CONTENTS (Continued)

TABLE OF CONTENTS (Continued)

APPENDICES

STANDARD FOR THE DESIGN AND FABRICATION OF GROUND SUPPORT EQUIPMENT

1. SCOPE

1.1 Purpose

The purpose of this Standard is to establish requirements and guidance for design and fabrication of ground support equipment (GSE) to assist National Aeronautics and Space Administration (NASA) space flight programs/projects in providing robust, safe, reliable, maintainable, supportable, and cost-effective GSE.

1.2 Applicability

a. This Standard is applicable to NASA space flight programs and its use is at the discretion of each program. This Standard establishes a set of GSE design requirements for NASA programs and projects. This Standard is intended for use in establishing uniform engineering practices and methods and ensuring that essential requirements are included in the design, procurement, and fabrication of GSE used to support the operations of receiving, storage, transportation, handling, assembly, inspection, test, checkout, service, launch, and recovery of space vehicles and payloads at NASA's launch, landing, or retrieval locations. GSE must also meet the requirements for the location of its intended use (e.g. AFSPCMAN 91-710, Range Safety User Requirements Manual, for the Eastern and Western Test Ranges).

b. This Standard is approved for use by NASA Headquarters and NASA Centers, including Component Facilities and Technical and Service Support Centers, and may be cited in contract, program, and other Agency documents as a technical requirement. This Standard may also apply to the Jet Propulsion Laboratory or to other contractors, grant recipients, or parties to agreements only to the extent specified or referenced in their contracts, grants, or agreements.

c. Requirements are numbered and indicated by the word "shall." Explanatory or guidance text is indicated in italics beginning in section 4.

d. Programs, projects, and elements are responsible for flowing requirements down to contractors, subcontractors, and suppliers of components at the lowest level.

e. Programs, in conjunction with Engineering and Safety and Mission Assurance (S&MA) organizations, have the responsibility to do the following:

 (1) Determine categories or types of GSE based on risk and any additional requirements resulting from these categories or types.

APPROVED FOR PUBLIC RELEASE – DISTRIBUTION IS UNLIMITED

(2) Establish and define where this Standard will be applied and when such boundaries require clarification.

(3) Establish program- and project-specific requirements.

(4) Establish minimum configuration management systems to retain the required documentation.

f. This Standard does not apply to the following unless specified by a program, project or Center:

(1) Equipment that is used solely during the manufacturing of flight hardware.

(2) Ground support systems (GSS) that interface with GSE.

(3) Facilities.

(4) Tools.

1.3   Tailoring

Tailoring of this Standard for application to a specific program or project shall be formally documented as part of program or project requirements and approved by the Technical Authority.

2.   APPLICABLE DOCUMENTS

2.1   General

The documents listed in this section contain provisions that constitute requirements of this Standard as cited in the text.

2.1.1   The latest issuances of cited documents shall apply unless specific versions are designated.

2.1.2   Non-use of specific versions as designated shall be approved by the responsible Technical Authority.

The applicable documents are accessible via the NASA Standards and Technical Assistance Resource Tool at https://standards.nasa.gov or may be obtained directly from the Standards Developing Organizations or other document distributors.

2.2   Government Documents

Department of Defense

MIL-DTL-16878	Wire, Electrical, Insulated, General Specification for
MIL-DTL-22992	Connectors, Plugs and Receptacles, Electrical, Waterproof, Quick Disconnect, Heavy Duty Type, General Specification for
MIL-DTL-24308	Connectors, Electric, Rectangular, Nonenvironmental, Miniature, Polarized Shell, Rack and Panel, General Specification for
MIL-DTL-38999	Connectors, Electrical, Circular, Miniature, High Density, Quick Disconnect (Bayonet, Threaded and Breech Coupling), Environment Resistant, Removable Crimp and Hermetic Solder Contacts, General Specification for
MIL-HDBK-6870	Inspection Program Requirements, Nondestructive for Aircraft and Missile Materials and Parts
MIL-PRF-39012	Connectors, Coaxial, Radio Frequency, General Specification for
MIL-STD-130	Identification Marking of U.S. Military Property
MIL-STD-461	Requirements for the Control of Electromagnetic Interference Characteristics of Subsystems and Equipment
MIL-STD-810	Environmental Engineering Considerations and Laboratory Tests
MIL-STD-889	Dissimilar Metals

Federal (FED)

29 CFR 1910	Occupational Safety and Health Standards
29 CFR 1910.1200	Hazard Communication and Center Requirements

49 CFR 171 thru 180	Hazardous Materials Regulations (Department of Transportation)
HF-STD-001	Human Factors Design Standard (HFDS) (Federal Aviation Administration)

NASA

NASA-HDBK-1001	Terrestrial Environment (Climatic) Criteria Handbook for Use in Aerospace Vehicle Development
NASA-STD-4003	Electrical Bonding for NASA Launch Vehicles, Spacecraft, Payloads, and Flight Equipment
NASA-STD-5008	Protective Coating of Carbon Steel, Stainless Steel, and Aluminum on Launch Structures, Facilities, and Ground Support Equipment
NASA-STD-5009	Nondestructive Evaluation Requirements for Fracture-Critical Metallic Components
NASA-STD-6001	Flammability, Offgassing, and Compatibility Requirements and Test Procedures
NASA-STD-6002	Applying Data Matrix Identification Symbols on Aerospace Parts
NASA-STD-8719.9	Standard for Lifting Devices and Equipment
NASA-STD-8719.12	Safety Standard for Explosives, Propellants, and Pyrotechnics
NASA-STD-8719.17	NASA Requirements for Ground-Based Pressure Vessels and Pressure Systems (PVS)
NASA-STD-8739.1	Workmanship Standard for Polymeric Application on Electronic Assemblies
NASA-STD-8739.4	Crimping, Interconnecting Cables, Harnesses, and Wiring
NASA-STD-8739.5	Fiber Optic Terminations, Cable Assemblies, and Installation
NPD 8730.1	Metrology and Calibration

NPR 2810.1	Security of Information Technology
NPR 6000.1	Requirements for Packaging, Handling, and Transportation for Aeronautical and Space Systems, Equipment, and Associated Components
NPR 7150.2	NASA Software Engineering Requirements
NPR 8715.3	NASA General Safety Program Requirements

NASA Centers

George C. Marshall Space Flight Center (MSFC)

| MSFC-STD-3029 | Guidelines for the Selection of Metallic Materials for Stress Corrosion Cracking Resistance in Sodium Chloride Environments |

John F. Kennedy Space Center (KSC)

KSC-GP-425	Fluid Fitting Engineering Standards
KSC-STD-E-0004	Pneumatic and Hydraulic Mechanical Components, Electrical Design, Standard for
MMA-1985-79	Standard Test Method for Evaluating Triboelectric Charge Generation and Decay

2.3   Non-Government Documents

Aerospace Industries Association (AIA)/National Aerospace Standards (NAS)

| AIA/NAS 410 | NAS Certification & Qualification of Nondestructive Test Personnel |

Aluminum Association (AA)

| ADM | Aluminum Design Manual |

American Institute of Steel Construction (AISC)

| AISC 325 | Steel Construction Manual |

APPROVED FOR PUBLIC RELEASE – DISTRIBUTION IS UNLIMITED

American Society of Civil Engineers (ASCE)

ASCE-7   Minimum Design Loads for Buildings and Other Structures

American Society of Heating, Refrigerating and Air-Conditioning Engineers (ASHRAE)

ASHRAE   Handbook—Fundamentals, Chapter 39

American Society for Testing and Materials (ASTM)

ASTM A123   Standard Specification for Zinc (Hot-Dip Galvanized) Coatings on Iron and Steel Products

ASTM A153   Standard Specification for Zinc Coating (Hot-Dip) on Iron and Steel Hardware

ASTM A249   Standard Specification for Welded Austenitic Steel Boiler, Superheater, Heat- Exchanger, and Condenser Tubes

ASTM A269   Standard Specification for Seamless and Welded Austenitic Stainless Steel Tubing for General Service

ASTM A312   Standard Specification for Seamless, Welded, and Heavily Cold Worked Austenitic Stainless Steel Pipes

ASTM A380   Standard Practice for Cleaning, Descaling, and Passivation of Stainless Steel Parts, Equipment, and Systems

ASTM A653   Standard Specification for Steel Sheet, Zinc-Coated (Galvanized) or Zinc-Iron Alloy-Coated (Galvannealed) by the Hot-Dip Process

ASTM A780   Standard Practice for Repair of Damaged and Uncoated Areas of Hot-Dip Galvanized Coatings

ASTM A1016   Standard Specification for General Requirements for Ferritic Alloy Steel, Austenitic Alloy Steel, and Stainless Steel Tubes

ASTM B241   Standard Specification for Aluminum and Aluminum-Alloy Seamless Pipe and Seamless Extruded Tube

ASTM D4174   Standard Practice for Cleaning, Flushing, and Purification of Petroleum Fluid Hydraulic Systems

ASTM D7194	Standard Specification for Aerospace Parts Machined from Polychlorotrifluoroethylene (PCTFE)
ASTM E426	Standard Practice for Electromagnetic (Eddy-Current) Examination of Seamless and Welded Tubular Products, Austenitic Stainless Steel and Similar Alloys
ASTM E1548	Standard Practice for Preparation of Aerospace Contamination Control Plans
ASTM E2217	Standard Practice for Design and Construction of Aerospace Cleanrooms and Contamination Controlled Areas

American Society of Mechanical Engineers (ASME)

ASME Y14.100	Engineering Drawing Practices
ASME B16.5	Pipe Flanges and Flanged Fittings NPS 1/2 Through NPS 24 Metric/Inch Standard
ASME B16.9	Factory-Made Wrought Buttwelding Fittings
ASME B31.3	Process Piping
ASME Boiler and Pressure Vessel Code, Section VIII, Divisions 1, 2, and 3	Rules for Construction of Pressure Vessels
ASME Boiler and Pressure Vessel Code, Section X	Fiber-Reinforced Plastic Pressure Vessels

American Welding Society (AWS)

AWS C3.2M/C3.2	Standard Method for Evaluating the Strength of Brazed Joints
AWS C3.4	Specification for Torch Brazing
AWS C3.5	Specification for Induction Brazing
AWS C3.6	Specification for Furnace Brazing

| AWS C3.7M/C3.7 | Specification for Aluminum Brazing |
| AWS D17.1 | Specification for Fusion Welding for Aerospace Applications |

Electronic Industries Association (EIA)/Electronic Components Association (ECA)

| EIA/ECA 310 | Cabinets, Racks, Panels, and Associated Equipment |

International Electrotechnical Commission (IEC)

| IEC 60807 | Rectangular Connectors for Frequencies Below 3 MHz |

International Organization for Standardization (ISO)

| ISO 14952 | Space Systems Surface Cleanliness of Fluid Systems |

International Telecommunication Union (ITU)

| ITU-T G.651 | Characteristics of a 50/125 Micrometer Multimode Graded Index Optical Fibre Cable |
| ITU-T G.652 | Characteristics of a Single-Mode Optical Fibre and Cable |

IPC – Association Connecting Electronics' Industries

IPC-2221	Generic Standard on Printed Board Design
IPC-2222	Sectional Design Standard for Rigid Organic Printed Boards
IPC-2223	Sectional Design Standard for Flexible Printed Boards
IPC-2252	Design Guide for RF/Microwave Circuit Boards
IPC-6011	Generic Performance Specification for Printed Boards,
IPC-6012	Qualification and Performance Specification for Rigid Printed Boards
IPC-6013	Qualification and Performance Specification for Flexible Printed Boards
IPC-6018	Microwave End Product Board Inspection and Test
IPC J-STD-001	Requirements for Soldered Electrical and Electronic Assemblies

IPC J-STD-001ES	Space Applications Electronic Hardware Addendum to J-STD-001 Requirements for Soldered Electrical and Electronic Assemblies
IPC-WHMA-A-620	Requirements and Acceptance for Cable and Wire Harness Assemblies

National Electrical Manufacturers Association (NEMA)

NEMA ICS 2	Industrial Control and Systems: Controllers, Contactors, and Overload Relays Rated 600 Volts
NEMA MG 1	Motors and Generators

National Fire Protection Association (NFPA)

NFPA 55	Compressed Gases and Cryogenic Fluids Code
NFPA 59A	Standard for the Production, Storage, and Handling of Liquefied Natural Gas (LNG)
NFPA 70	National Electrical Code
NFPA 70E	Standard for Electrical Safety in the Workplace
NFPA 496	Standard for Purged and Pressurized Enclosures for Electrical Equipment

Sheet Metal and Air Conditioning Contractors' National Association (SMACNA)

SMACNA1958	HVAC Systems - Duct Design

Society of Automotive Engineers (SAE)

SAE AIR 4071	Lubricants for Oxygen Use
SAE AMS 2175	Castings, Classification and Inspection of
SAE AMS 2403	Plating, Nickel General Purpose
SAE AMS 2404	Plating, Electroless Nickel
SAE AMS 2423	Plating, Nickel Hard Deposit

SAE AMS 2488	Anodic Treatment – Titanium and Titanium Alloys Solution pH 13 or Higher
SAE AMS 2759	Heat Treatment of Steel Parts, General Requirements
SAE AMS 2759/9	Hydrogen Embrittlement Relief (Baking) of Steel Parts
SAE AMS 2770	Heat Treatment of Wrought Aluminum Alloy Parts
SAE AMS 2771	Heat Treatment of Aluminum Alloy Castings
SAE AMS 2772	Heat Treatment of Aluminum Alloy Raw Materials
SAE AMS 2774	Heat Treatment Wrought Nickel Alloy and Cobalt Alloy Parts
SAE AMS-H-6875	Heat Treatment of Steel Raw Materials
SAE AMS-H-81200	Heat Treatment of Titanium and Titanium Alloys
SAE AS 22759	Wire, Electrical, Fluoropolymer-Insulated, Copper or Copper Alloy
SAE AS 50151	Connectors, Electrical, Circular Threaded, AN Type, General Specification for
SAE AS 50861	Wire, Electric, Polyvinyl Chloride Insulated, Copper or Copper Alloy

Telcordia

GR-20	Generic Requirements for Optical Fiber and Optical Fiber Cable
GR-409	Generic Requirements for Premises Fiber Optic Cable

2.4   Order of Precedence

This Standard establishes requirements and guidance for design and fabrication of GSE to assist NASA space flight programs/projects in providing robust, safe, reliable, maintainable, supportable, and cost-effective GSE but does not supersede nor waive established Agency requirements found in other documentation.

2.4.1   Conflicts between this Standard and other requirements documents shall be resolved by the responsible Technical Authority.

3. ACRONYMS AND DEFINITIONS

3.1 Acronyms and Abbreviations

°C	degrees Celsius
°F	degrees Fahrenheit
AA	Aluminum Association
ADM	Aluminum Design Manual
AFSPCMAN	Air Force Space Command Manual
AIA	Aerospace Industries Association
AIAA	American Institute of Aeronautics and Astronautics
AISC	American Institute of Steel Construction
AMS	Aerospace Material Specification
AS	Aerospace Standard
ASCE	American Society of Civil Engineers
ASHRAE	American Society of Heating, Refrigerating and Air-Conditioning Engineers
ASME	American Society of Mechanical Engineers
ASTM	American Society for Testing and Materials
AWS	American Welding Society
CFR	Code of Federal Regulations
CGA	Compressed Gas Association
COTS	commercial off-the-shelf
DTL	detail
ECA	Electronic Components, Assemblies & Materials Association
ECS	Environmental Control System
EEE	electrical, electronic, and electromechanical
e.g.	for example
EIA	Electronic Industries Association
EMC	electromagnetic compatibility
EMI	electromagnetic interference
ESD	electrostatic discharge
ETFE	ethylene tetrafluoroethylene
FOD	foreign-object debris
GO_2	gaseous oxygen
GR	generic requirement
GSE	ground support equipment
GSS	ground support system
HDBK	handbook
i.e.	that is
IEC	International Electrotechnical Commission
IEEE	Institute of Electrical and Electronic Engineers
in	inch
IPC	Association Connecting Electronics Industries

ISO	International Organization for Standardization
ITU	International Telecommunication Union
JSC	Lyndon B. Johnson Space Center
kPa	kilopascal
KSC	John F. Kennedy Space Center
ksi	one thousand pounds per square inch
KTI	Kennedy Technical Instruction
LH_2	liquid hydrogen
LNG	liquefied natural gas
LO_2	liquid oxygen
M&P	materials and processes
	Materials and Processes organization
MAPTIS	Materials and Processes Technical Information System
MHz	megahertz
MIL	military
mm	millimeter
MMA	Malfunction/Materials Analysis
MMH	monomethylhydrazine
MMPDS	Metallic Materials Properties Development and Standardization
MNL	manual
MPa	megapascal
MSFC	George C. Marshall Space Flight Center
NAFPI	National Aerospace FOD Prevention Inc.
NAS	National Aerospace Standards
NASA	National Aeronautics and Space Administration
NDE	nondestructive evaluation
NEMA	National Electrical Manufacturers Association
NFPA	National Fire Protection Association
NPD	NASA Policy Directive
NPR	NASA Procedural Requirements
NPT	National Pipe Thread Tapered Thread
PC	printed circuit
PCA	printed circuit assembly
PCB	printed circuit board
PCTFE	polychlorotrifluoroethylene
PHE	propellant handler's ensemble
PPE	personnel protective equipment
ppm	parts per million
PRF	performance specification
PQR	procedure qualification record
psia	pound per square inch absolute
psig	pounds per square inch gauge
PVS	pressure vessels and pressure systems
PTFE	polytetrafluoroethylene

QD	quick disconnect
RF	radio frequency
RH	relative humidity
S&MA	Safety and Mission Assurance
SAE	Society of Automotive Engineers
SCC	stress corrosion cracking
SMACNA	Sheet Metal and Air Conditioning Contractors' National Association
SPEC	specification
SSP	Space Shuttle Program
STD	standard
TM	technical memorandum
TP	technical procedure
UNS	Unified Numbering System
UTS	ultimate tensile strength
V/m	volts per meter
vs.	versus
WPQ	welding operator performance qualification
WPS	welding procedure specification

3.2   Definitions

Commercial Off-The-Shelf (COTS): Equipment, including hardware and associated software, that is commercially available from the industrial inventory at the time of purchase.

Rationale: Commercial items or components should be used when they satisfy the GSE function and will not degrade the safety or reliability of the ground or flight system. Requirements should be specified in terms of functionality or performance rather than design.

Critical GSE: GSE whose loss of function or improper performance could result in serious personnel injury, damage to flight hardware, loss of mission, or major damage to a significant ground asset.

Design Life: The operational life of equipment (to include storage life, installed life in a nonoperating mode, and operational service life), after which the equipment will be replaced or recertified. It is the responsibility of the program/project to determine recertification requirements, which may include refurbishment, analysis, or test.

Factor of Safety (Safety Factor): A constant that has been defined for yield and ultimate design criteria and that is the ratio of the yield or ultimate design loads to the limit load (the maximum allowable design load). If the factor of safety is defined in terms of stress, it is the ratio of the ultimate or yield stress to the maximum design stress. In fatigue design, it is the ratio of the calculated fatigue life to the design life. This Standard specifies the minimum factor of safety for GSE for specific structural applications (e.g., pressure vessels, threaded fasteners, and aluminum structures).

Rationale: This definition is consistent between ground and flight hardware. This definition is inherently load-based. It reduces to the traditional stress-based definition in the simplest case. The basis for the factor of safety should be stated in the engineering calculations.

Flight Hardware: Hardware intended for launch, including boosters, engines, payloads, and manned or unmanned components.

Fracture-Critical: Classification of hardware where a crack could lead to a failure that results in serious personnel injury, damage to flight hardware, loss of mission, or major damage to a significant ground asset.

Rationale: The term fracture-critical for GSE is used only to determine requirements as specified in this Standard and does not imply additional fracture-critical requirements normally applied to flight hardware.

Ground Support Equipment (GSE): Nonflight equipment, systems, or devices specifically designed and developed for a physical or direct functional interface with flight hardware.

Rationale: GSE is equipment whose failure could propagate into flight hardware or equipment that could prevent the failure of GSS and facility systems from propagating into flight hardware. Equipment used during the manufacturing of flight hardware is not considered to be GSE. Each program defines when manufacturing ends and processing of the flight hardware begins. If manufacturing equipment is to be used after flight hardware processing begins, it must be designed to meet GSE requirements. GSE does not include tools that are designed for general use and not specifically for use on flight hardware.

Ground Support Systems (GSS): Infrastructure and equipment (portable or fixed) that provides functional or physical support to GSE and does not directly interface with flight hardware, although it may supply commodities, power, or data that eventually reaches the flight hardware after being conditioned or controlled by GSE.

Rationale: Design standards for GSS may be similar to or, at the discretion of the program/project, be identical to the design standards for GSE. Protective features designed into the GSE prevent failures from propagating to flight hardware.

Inadvertent Operator Action: A single error by the operator of a GSE control function on a configured system.

Rationale: Operator errors include actions performed at an inappropriate time or to an incorrect degree such as operating a switch, button, valve, knob, lever, or other design feature of the system requiring operator input. This does not include cross-connecting fluid lines/electrical cables or performing unsafe acts not related to operation of the system.

Safe Working Load: The maximum assigned load the device or equipment can operationally handle and maintain. This value is marked on the device indicating maximum working capacity. This is also the load referred to as "rated load" or "working load limit." If the device has never been downrated or uprated, this also is the "manufacturer's rated load."

Tools: Equipment designed for general use in a variety of applications. Tools are calibrated, when necessary, in accordance with industrial standards.

Rationale: Tools are not designed to specifically interface with flight hardware, nor are they designed to perform a function specific to flight hardware. Their design and general use in industry includes a variety of applications that may be required on flight hardware or GSE. Tools are intended for use by trained technicians and facilitate manual operations, such as torquing fasteners, cutting wire, checking electrical continuity, and verifying surface clearances. Examples of tools include torque wrenches, crow's feet, voltmeters, go/no-go gages, screwdrivers, wire cutters, and pliers.

4. GENERAL REQUIREMENTS

4.1 Commercial Off-The-Shelf (COTS) Equipment

In order to meet customer requirements, individual system and equipment design projects may need criteria that are more stringent than those specified herein. In such cases, these criteria should be determined by the responsible design organization in consultation with its customers (e.g., users and operators).

Each program/project has the responsibility to define its own policy for the acceptance of COTS equipment in GSE.

When a program/project approves the use of COTS equipment in GSE, the following design requirements apply:

     a. COTS equipment shall be evaluated for acceptability from a materials and processes (M&P) standpoint (see section 6).

     b. Qualification tests and inspections shall be indicated in the engineering documentation.

     c. For COTS used as critical GSE, documentation shall be provided as evidence that the requirements of this document have been met.

     d. Modifications to COTS shall be performed in accordance with the design and fabrication requirements in this document.

     e. COTS incorporated into GSE shall be selected for use within the limits of the manufacturer's specified ratings (e.g., environmental, mechanical, electrical, electromagnetic compatibility, etc.).

COTS equipment should be used to the maximum extent possible when (1) it satisfies the intended function; (2) it will not degrade the safety or reliability of the flight or ground system; and (3) it provides a cost savings that exceeds possible cost increases due to unique maintenance or logistics requirements, modifications, or an increase in the complexity of the interfacing equipment. Vendor or contractor documentation and supporting test data should be incorporated into system control documents.

4.2 System Requirements

4.2.1 Performance Requirements

In addition to performance requirements, GSE should be designed for ease of production, manufacture, construction, and inspection. GSE should be designed to minimize the complexity and frequency of maintenance. Close manufacturing tolerances should be avoided unless required by design and performance.

4.2.1.1 Design Life

GSE shall be designed for the operational life specified by program or mission requirements.

Engineering documentation should specify maintenance requirements to meet design life.

4.2.1.2 Contamination from GSE

GSE shall not cause degradation of or contaminate flight systems or ground systems while it is being used, checked out, serviced, or otherwise handled.

4.2.1.3 GSE Design for Access

GSE design shall include access provisions for handling, servicing, calibration, maintenance, repair, and replacement of components and limited-life items.

GSE design should provide for ease of operation, maintenance, servicing, cleaning, and inspection of hardware and software. GSE fault detection and isolation should be considered based on criticality and cost of failures.

4.2.1.4 Interfaces

a. GSE shall meet the requirements of all interfaces with flight hardware or software as specified in interface control documentation.

Interfaces should be verified by test and/or analysis. GSE should be designed to reduce or eliminate the potential for a mating device (fluid, mechanical, or electrical) to be connected to the wrong interface with flight hardware.

b. GSE shall be compatible with all facility interfaces.

An assessment may be required to determine if it is more cost-effective to modify the facility interface or design the GSE to meet the existing facility interface.

4.2.2 Heritage GSE

Heritage GSE is hardware/software that has been fabricated and delivered to a previous program. GSE designs that were not fabricated are not considered Heritage GSE and must meet the requirements of this Standard.

Existing GSE that was verified to meet the GSE design and fabrication requirements of a previous NASA program shall be considered acceptable for use without further verification to the requirements of this Standard, if the following conditions are met:

    a. Approval is obtained from the responsible program/project office.

    b. Waivers to GSE design and fabrication requirements from the previous NASA program are evaluated and approved by the responsible Technical Authority against the equivalent requirements in this Standard.

    c. Existing GSE is verified to meet the performance requirements, interfaces, operational limits, and environments for its intended application.

GSE design and fabrication requirements for previous NASA programs are defined in SW-E-0002, Ground Support Equipment General Design Requirements: Space Shuttle; KSC-DE-512-SM, Facility, System, and Equipment General Design Requirements; SSP 50004, Ground Support Equipment Design Requirements: International Space Station; or a previous version of this Standard.

4.2.3 Reliability

4.2.3.1 Redundancy

When redundancy is required to satisfy failure tolerance:

    a. Redundant systems, subsystems, or components shall be designed such that the failure of one will not prevent the other from performing its intended function.

Physically orienting or separating redundant items should be considered to prevent propagation of failures (e.g., routing electrical cables in separate cable trays).

    b. Redundant systems, subsystems, or components shall be designed such that common-cause failures (e.g., contamination, common power source, etc.) do not eliminate redundancy.

    c. Redundant systems shall be designed such that each independent path can be tested without invalidating the system configuration.

4.2.3.2 Failure Tolerance

    a. GSE (except primary structure and pressure vessels, piping, and tubing in rupture mode) shall be designed such that no single failure or inadvertent operator action results in equipment or facility damage or personal injury.

Failure tolerance is not possible for primary structure and pressure vessels/tubing in rupture mode. Safety of these systems is achieved by application of design requirements contained in this standard.

GSE may be designed to terminate operations autonomously after the first failure or inadvertent operator action and in time to preclude any scenario that results in equipment or facility damage or personal injury. This approach is consistent with the historical use of the term "fail-safe" by the GSE design community.

b. GSE shall be designed to sustain a failure or inadvertent operator action and still perform its basic function when necessary to meet either safety or operational requirements.

Purges, ground special power, and launch release systems are examples of systems that interface with other systems whose loss would propagate failures. This approach is consistent with the historical use of the term "fail-operational" by the GSE design community.

4.2.3.3 Failure Propagation

GSE shall be designed to prevent failures from propagating to flight systems.

The design of GSE should consider how flight hardware/software failures could propagate through the GSE and affect other flight systems.

4.2.4 Environmental Conditions

4.2.4.1 Natural Environment

GSE used or stored in an uncontrolled exterior environment shall be designed to function after exposure to the natural environment at its respective geographical location as specified in NASA-HDBK-1001, Terrestrial Environment (Climatic) Criteria Handbook for Use in Aerospace Vehicle Development.

Specifications in NASA-HDBK-1001 may be tailored to reflect program-defined risk and exposure times, including operation within the launch commit criteria of the vehicle.

4.2.4.2 Launch-Induced Environment

a. GSE required to function during or after exposure to the launch-induced environment shall be designed to withstand the environment defined in program documents.

b. GSE not required to function after exposure to the launch-induced environment shall not cause damage or create a hazard to flight hardware, facilities, other GSE, personnel, or the environment.

4.2.4.3   Controlled Interior Environment

GSE designed to function within a controlled interior environment shall be designed to the following temperature and humidity requirements:

    a. Nominal temperature: +15 °C (60 °F) to +27 °C (80 °F).

    b. Extreme temperature: +11 °C (52 °F) to +15 °C (60 °F) or +27 °C (80 °F) to +40 °C (105 °F) for a maximum of 1 hour.

    c. Humidity: nominal 55 percent, within a range of 30 percent to 70 percent.

4.2.4.4   Controlled Clean Environment

Clean rooms and associated controlled environments designated as GSE shall be designed in accordance with ASTM E2217, Standard Practice for Design and Construction of Aerospace Cleanrooms and Contamination Controlled Areas.

4.2.4.5   Uncontrolled Interior Environment

GSE used in an uncontrolled interior environment shall be designed to meet the most severe exterior environmental conditions for humidity and temperature anticipated at the respective geographical locations, as defined in NASA-HDBK-1001.

Some uncontrolled interior environments can exceed the most severe exterior environment, i.e., an enclosed trailer in a hot (tropical or desert) exterior environment.

4.2.4.6   Environmental Test Methods

Environmental methods and conditions required for testing and qualification of GSE components shall be in accordance with MIL-STD-810, Environmental Engineering Considerations and Laboratory Tests.

KSC-STD-164, Environmental Test Methods for Ground Support Equipment, Standard for, should be used as a guide to standardize practices.

4.2.4.7   Seismic Environment

GSE used in Zones 3 or 4 as defined in NASA-HDBK-1001 shall be designed to resist the effects of a seismic event using the criteria and guidelines in ASCE-7, Minimum Design Loads for Buildings and Other Structures, Chapter 15, Seismic Design Requirements for Non-Building Structures.

4.3 Documentation

4.3.1 Drawings and Specifications

Drawings and specifications required for the fabrication, construction, installation, modification, test, operation, maintenance, sustaining, and use of GSE shall be prepared in accordance with drawing practices equal to or more stringent than the engineering drawing practices of ASME Y14.100, Engineering Drawing Practices.

4.3.2 Technical Documentation

Technical documentation (e.g., manuals and reports) shall be prepared and delivered to the user of the GSE.

Documentation should include applicable deliverables contained in Appendix B.

4.4 Logistics

GSE design documentation shall identify spare parts, components, materials, and items required to support construction, fabrication, installation, activation, test, verification, and operation.

4.5 Limited-Life Items

a. Limited-life items shall be controlled from the date of their manufacture through their period of operational use, including the time they are in storage.

b. The status of limited-life items shall be documented.

c. GSE design documentation shall identify limited-life items and provide instructions for replacement of those parts, components, and materials.

Use of items with a projected lifetime that is less than the design life of the GSE for which the items are intended should be avoided whenever possible. Elapsed time or cycle indicators should be employed to accumulate operational time or cycles for limited-life items. The age of items that are installed in a non-operating mode should also be tracked.

4.6 Quality Assurance

Documentation shall be provided by the design organization to verify compliance with this Standard (reference Appendix B).

Examinations and tests are recommended to verify that all requirements of this Standard have been achieved. This may include the following:

- *Tests and analyses of the performance and reliability requirements.*

- *Measurement or comparison of specified physical characteristics.*

- *Verification, with specific criteria, of workmanship.*

- *Test and inspection methods for ensuring compliance, including environmental.*

- *Conditions for performance.*

4.6.1   Quality Requirements Definition

Quality requirements shall be defined on the engineering drawings or in other technical documents that are included in the design, fabrication, or installation contract.

The design documentation should include any special processes, testing, and inspections to be conducted.

4.6.2   Testing

a.   Testing requirements shall be specified in engineering documentation.

b.   Critical components shall undergo qualification testing to verify compliance with the applicable specifications and the ability to perform required design functions in the intended environment.

4.6.2.1   Load Test

a.   Structural GSE (access platforms, workstands, etc.) shall be load tested to a minimum of 125 percent of the design or working load.

b.   Lifting devices and equipment shall be load tested in accordance with NASA-STD-8719.9, Standard for Lifting Devices and Equipment.

4.6.2.2   Metrology and Calibration

Metrology and calibration of all test equipment and tools used in support of fabrication, assembly, installation, and test of GSE shall be in accordance with NPD 8730.1, Metrology and Calibration.

4.7 Packaging, Handling, and Transportation

Packaging, transporting, shipping, and handling shall be in accordance with NPR 6000.1, Requirements for Packaging, Handling, and Transportation for Aeronautical and Space Systems, Equipment, and Associated Components.

GSE should be designed so that it can be transported by ground, air, or sea using commercially available methods. Containers should be designed to accommodate all planned transportation, handling, and storage methods. ESD-sensitive components and assemblies should be handled using practices in accordance with ANSI/ESD S20.20, ESD Association Standard for the Development of an Electrostatic Discharge Control Program for - Protection of Electrical and Electronic Parts, Assemblies and Equipment (Excluding Electrically Initiated Explosive Devices).

5. DESIGN AND FABRICATION REQUIREMENTS

5.1 Structural Design

5.1.1 GSE Structures and Equipment

5.1.1.1 Steel Structures

The design of steel structures fabricated from conventional rolled shapes shall be in accordance with the AISC 325, Steel Construction Manual, using the Allowable Stress Design method.

5.1.1.2 Aluminum Structures

The design of aluminum structures fabricated from conventional rolled or extruded shapes shall be in accordance with Aluminum Association ADM, Aluminum Design Manual.

5.1.2 Factor of Safety

a. The following minimum factors of safety shall be used for support structures (excluding lifting devices and equipment, pressure systems, preload in threaded fasteners, and springs) when not otherwise specified.

(1) GSE structures shall be designed to a minimum factor of safety of 2 against deformation or yielding that impairs the function of the part and a minimum factor of safety of 3 against collapsing, buckling, exceeding the ultimate load, or failing to support the design load.

(2) For brittle materials (less than 5 percent elongation-to-failure) and fiber-reinforced polymers, the factor of safety shall be 5 against exceeding the ultimate tensile strength (UTS).

b. Threaded fasteners shall meet the factors of safety contained in (a) when considering only the operating load, not including preload which is covered in section 5.2.8

c. Springs shall be designed to a minimum factor of safety of 1.4 against exceeding the UTS.

Due to the nature of their design, springs do not meet the factors of safety specified for other engineering applications and should be designed in accordance with the Handbook of Spring Design from the Spring Manufacturers Institute.

d. Structures exposed to cyclic loads shall be designed for a minimum service life factor of 4 applied to the design life for fatigue assessments.

5.1.3 Structural Design Loads

Structural design loads shall be specified in the design documentation.

The design should address typical load cases such as the loads created by the assembly, transportation, and operations processes; wind conditions; the structure's lateral stability; and seismic events.

5.2 Mechanical Design

5.2.1 Pneumatics

a. The design of GSE used for pneumatic servicing, excluding shop air below 1.14 MPa (165 psia), shall be in accordance with ASME Boiler Pressure Vessel Code, Section VIII and ASME B31.3, Process Piping, with 100 percent radiography of welded joints.

b. Hoses shall consist of a polytetrafluoroethylene or convoluted 300-series stainless steel inner tube reinforced with a stainless steel wire construction of braid or spiral wrap.

c. Flex hose installations designed for 1.14 MPa (165 psia) or greater shall incorporate hose restraints.

d. For pressurized air systems greater than 1.83 MPa (265 psia) and for all oxygen systems, component bodies and other pressure containing parts shall meet section 6.3.1.3.2.

e. Hydrogen systems shall conform to the requirements of NFPA 55, Compressed Gases and Cryogenic Fluids Code.

5.2.1.1   Breathing Air

Breathing air systems shall conform to 29 CFR 1910, Occupational Safety and Health Standards, Subparts H, I, and M.

KSC-STD-Z-0008, Standard for Design of Ground Life Support Systems and Equipment should be used as a guide to standardize design practices.

5.2.1.2   Interfaces to Hypergol Systems

a.   Each pneumatic line that interfaces with a hypergol system shall have a hand-operated shut-off valve upstream of a spring-loaded poppet-type check valve to permit positive shutoff of the pneumatic supply and prevent backflow through the pneumatic line.

b.   Each pneumatic line that interfaces with a hypergol system shall be downstream of a pneumatic supply pressure regulator that serves only a fuel or oxidizer system to prevent mixing of combustible vapors.

c.   A sampling port shall be provided upstream and downstream of each regulator to permit periodic sampling of the pneumatic fluid for hypergol contamination.

d.   A pressure gauge shall be provided downstream of each check valve to indicate the pressure in the hypergol system.

Gauge calibration ports should be designed to limit potential contaminated gas impingement on personnel.

KSC-STD-Z-0005, Design of Pneumatic Ground-Support Equipment, Standard for, should be used as a guide to standardize design practices.

5.2.2   Cryogenics

GSE used for all cryogenic servicing shall be in accordance with ASME Boiler and Pressure Vessel Code, Section VIII and ASME B31.3 (for Severe Cyclic Conditions).

a.   Liquid Hydrogen (LH_2) and Liquid Oxygen (LO_2) systems shall be designed in accordance with NFPA 55.

b.   The design of GSE used for cryogenic servicing with liquid methane shall be in accordance with NFPA 59A Standard for the Production, Storage, and Handling of Liquefied Natural Gas (LNG).

KSC-STD-Z-0009, Design of Cryogenic Ground Support Equipment Standard for, should be used as a guide for processes, methods, and practices applied to cryogenic GSE at Kennedy Space Center.

5.2.3 Hypergols

GSE used for hypergolic propellant and monopropellant servicing shall be in accordance with ASME Boiler Pressure Vessel Code, Section VIII, including lethal service requirements; and ASME B31.3, including Category M requirements (see 6.3.1.3.1 for material compatibility requirements).

    a. Hypergol system design shall preclude the chances of fuel and oxidizer intermixing by the following:

        (1) Color coding and marking of hypergol fuels and oxidizer servicing equipment.

        (2) Uniquely sizing or using a mechanically different type of connection for each commodity where close proximity creates a potential for cross-connection.

        (3) No interconnection between hypergol fuel and oxidizer piping systems.

    b. All hypergol servicing equipment shall be designed with spill containment, such as integrated drip pans or drainage to a collection/removal location.

    c. Pipe flanges shall be raised face with concentric serrations.

KSC-STD-Z-0006, Hypergolic Propellants Ground Support Equipment, Design of, Standard for, should be used as a guide to standardize design practices. National Pipe Thread Tapered Thread (NPT) connections should not be used in hypergolic servicing systems. The use of NPT should be restricted to those connections that require conversion from pipe to tube and must have ASME code safeguarding provisions, such as seal welds. GSE used for alternative propellants developed to replace hypergols should meet the requirements of this section.

5.2.4 Hydrocarbons

GSE used for servicing with hydrocarbon fuels shall be in accordance with ASME Boiler Pressure Vessel Code, Section VIII and ASME B31.3.

Low pressure (less than 15 psig) stationary ground system storage tanks may be designed in accordance with API Standard 620.

5.2.5 Hydraulics

GSE used for servicing hydraulic systems shall be in accordance with ASME Boiler Pressure Vessel Code, Section VIII and ASME B31.3 and cleaned to a level meeting or exceeding ASTM D4174, Standard Practice for Cleaning, Flushing, and Purification of Petroleum Fluid Hydraulic Systems.

GSE used for servicing hydraulic systems should be in accordance with National Fluid Power Association or International Organization for Standardization (ISO) standards. KSC-STD-Z-0005 should be used as a guide to standardize design practices.

5.2.6 Environmental Control Systems (ECS)

a. GSE used for an ECS shall be in accordance with ASHRAE Handbook— Fundamentals (Code and Standards), Chapter 39.

b. Gaseous nitrogen shall be isolated from the ECS air ducting system by using two valves in series in the GN_2 supply line and vented to exterior atmosphere between the valves.

c. If gaseous nitrogen supply is connected to the ECS duct system, ECS shall include the capability of monitoring the oxygen content in the ducting.

d. The design of ducting shall be in accordance with ASME B31.3 or SMACNA 1958, HVAC Systems - Duct Design.

Intakes to air distribution systems should be located remote from areas where normal toxic vapor venting occurs or where accidental spillage could result in contamination of breathing or vehicle air supplies and should provide toxic vapor protection of equipment located in areas where toxic vapors might accumulate. Pre-filters should be used in fresh air intakes and prior to primary filters to prevent excessive loading of the primary filter. Filters should be located immediately upstream of all interfaces where control of particulate matter is required for system performance.

5.2.7 Lifting Devices

The design and certification of lifting devices (cranes, crane girders, hoists, lifting slings, jacks, etc.) shall be in accordance with NASA-STD-8719.9.

5.2.8 Threaded Fasteners

a. The design of threaded fastener installation shall not exceed a preload of 70 percent of the yield stress on the net cross section of the fastener.

b. The installation criteria for structural bolts, such as specified in ASTM A325, Standard Specification for Structural Bolts, Steel, Heat Treated, 120/105 ksi Minimum Tensile Strength, and ASTM A490, Standard Specification for Structural Bolts, Alloy Steel, Heat Treated, 150 ksi Minimum Tensile Strength, shall be in accordance with AISC 325.

c. Installation criteria shall be documented on the fabrication, assembly, or installation drawing.

This section addresses the design of and maximum allowable preload for threaded fasteners. The design application may require a lower preload value due to actual applied loads, gaskets, seals, etc. Other sources for installation/torque criteria include MSFC-STD-486, Standard, Threaded Fasteners, Torque Limits for, and KSC-SPEC-Z-0008, Fabrication and Installation of Flared Tube Assemblies and Installation of Fittings and Fitting Assemblies, Specification for.

5.2.9 Tethering Provisions

a. GSE components that require temporary removal/installation (quick release pins, quick disconnect (QD) caps, etc.) during operational activities shall be tethered to or otherwise held captive by the equipment for which they are used.

b. GSE intended for use near flight hardware or elevated above personnel and/or flight hardware shall be designed and constructed with provisions for tethering.

5.2.10 Transportation Loads

Transportation equipment shall be designed so that loads imparted to flight hardware do not exceed 80 percent of the flight limit loads.

Transportation loads should be evaluated early in the design cycle since they may be the governing design load case. ASTM D4169, Performance Testing of Shipping Containers and Systems, should be used to determine transportation environments and test methods for the applicable transportation method.

5.2.11 Pressure Systems

All pressurized systems shall comply with NASA-STD-8719.17, NASA Requirements for Ground-Based Pressure Vessels and Pressurized Systems. Additional requirements for specific commodities and applications are listed in this Standard.

NASA-STD-8719.17 is required of all NASA designs and encompasses many of the specific requirements contained in this document. This Standard and the referenced standards herein are intended to provide additional requirements.

a.   Metallic pressure vessels for use in GSE shall be designed, constructed, tested, and stamped in accordance with ASME Boiler and Pressure Vessel Code, Section VIII, Division 1, 2, or 3, Rules for Construction of Pressure Vessels.

b.   Composite Overwound Pressure Vessels for use in GSE shall be designed, constructed, tested, and marked in accordance with ASME Boiler and Pressure Vessel Code, Section X, Fiber-Reinforced Plastic Pressure Vessels.

5.2.11.1   Code-Stamped Vessel Registration

All ASME code-stamped vessels shall be registered with the National Board of Boiler and Pressure Vessel Inspectors.

5.2.11.2   Transporting Hazardous Commodities

Pressure vessels used for transporting hazardous commodities shall meet the Department of Transportation requirements in 49 CFR 171 through 180, Hazardous Materials Regulations.

5.3   Electrical/Electronic Design

5.3.1   Electrical Control and Monitor Equipment

The design of electrical control and monitoring equipment shall provide features for safety of the commanded functions, essential monitoring to ensure proper performance, human factor considerations, and fault isolation for the equipment it is servicing.

KSC-STD-E-0001, Design of Electrical Control and Monitoring Systems, Equipment (GSE) and Panels, Standard for should be used as a guide to standardize design practices.

5.3.2   Electrical Design of Pneumatic and Hydraulic Components

The electrical design of pneumatic and hydraulic components shall be in accordance with KSC-STD-E-0004, Pneumatic and Hydraulic Mechanical Components, Electrical Design, Standard for.

5.3.3   Pyrotechnic Systems

The design of pyrotechnic GSE shall be in accordance with NASA-STD-8719.12, Safety Standard for Explosives, Propellants, and Pyrotechnics.

5.3.4   Electrical Power Systems

GSE electrical power systems shall be in accordance with NFPA 70, National Electric Code, and NFPA 70E, Standard for Electrical Safety in the Workplace.

Incorporation of batteries in the design of GSE should follow the recommended practices in the following documents:

a. IEEE 484, IEEE Recommended Practice for Installation Design and Installation of Vented Lead-Acid Batteries for Stationary Applications.

b. IEEE 1106, IEEE Recommended Practice for Installation, Maintenance, Testing, and Replacement of Vented Nickel-Cadmium Batteries for Stationary Applications.

c. IEEE 1187, IEEE Recommended Practice for Installation Design and Installation of Valve-Regulated Lead-Acid Storage Batteries for Stationary Applications.

d. IEEE 446, IEEE Recommended Practice for Emergency and Standby Power Systems for Industrial and Commercial Applications.

5.3.5 Bonding, Grounding, and Lightning Protection

a. Bonding and grounding shall be provided in accordance with NFPA 70 for connections within the GSE and for connections to facilities or GSS.

b. Bonding and grounding shall be provided in accordance with NASA-STD-4003, Electrical Bonding for NASA Launch Vehicles, Spacecraft, Payloads, and Flight Equipment, for connections to flight systems.

c. GSE not shielded by a lightning protection system shall meet its operational performance requirements after being subjected to the indirect effects of lightning defined as follows:

- Magnetic field rate of change @10 meters = 2200 amperes per meter per μsec.

- Electric field rate of change @ 10 meters = 680 kilo-volts per meter per μsec.

d. GSE located outdoors and shielded by a lightning protection system shall meet its operational performance requirements after being subjected to the indirect effects of lightning defined as follows:

- Magnetic field rate of change = 130 amperes per meter per μsec.

- Electric field rate of change = 40 kilo-volts per meter per μsec.

Indirect effects of lightning are electrical transients due to coupling of electromagnetic fields that may cause damage to or malfunction of electrical/electronic equipment. GSE should be located in areas where lightning protection is provided or designed to sustain a near-direct (within 10 meters) lightning strike. GSE located in some protected areas will experience very

APPROVED FOR PUBLIC RELEASE – DISTRIBUTION IS UNLIMITED

high induced voltages and current from nearby lightning strikes and must provide internal protection to assure each GSE item will perform as designed.

5.3.6 Electrostatic Discharge

GSE containing electronic components shall be designed to protect against a contact discharge of 8 kV.

5.3.7 Hazard Proofing

The design of electronic equipment and wiring for all voltages in hazardous locations shall be in accordance with NFPA 70, Article 500.

NFPA 70 defines hazardous locations and methods of hazard proofing including purging, intrinsically safe, and Class/Division ratings. KSC-STD-E-0002, Hazard Proofing of Electrically Energized Equipment, Standard for, should be used as a guide to standardize design practices.

5.3.8 Software

Software incorporated in the design of GSE shall meet the requirements of NPR 7150.2, NASA Software Engineering Requirements.

Software includes COTS software, government-off-the-shelf software, modified-off-the-shelf software, reused software, auto-generated code, embedded software, firmware, and open source software components.

5.4 Parts

GSE must meet Center and program/project parts policies and requirements established in accordance with NPD 8730.2, NASA Parts Policy.

5.4.1 GSE Fasteners

5.4.1.1 Reuse of Self-Locking Fasteners

a. The reuse of self-locking fasteners shall be permitted when the running torque before clamp-up remains between the maximum self-locking torque and the minimum breakaway torque.

Self-locking fasteners should be used wherever possible. The use of star lock washers should be avoided.

b. Fasteners used in corrosive environments and applications where condensation can occur shall be installed using a corrosion-resistant sealant while the sealant is still wet (wet installation).

5.4.1.2   Liquid-Locking Compounds

When liquid-locking compounds are used for fastener installation, engineering drawings shall specify a validated process for application.

Liquid-locking compounds should be selected in accordance with ASTM D5363, Standard Specification for Anaerobic Single-Component Adhesives (AN).

5.4.1.3   Critical GSE Fasteners

Critical GSE fasteners shall have lot traceability from the manufacturer to final installation.

Traceability applies to any single fastener if its failure will cause injury, loss of life, or damage to a flight or ground system.

5.4.2   Stainless Steel Tubing

All stainless steel tubing should be fabricated, tested, and installed in accordance with KSC-SPEC-Z-0008.

a. Austenitic stainless steel tubing shall be in accordance with ASTM A269, Standard Specification for Seamless and Welded Austenitic Stainless Steel Tubing for General Service, with the following specifications:

(1) Wall thickness tolerance of +10 percent/-0 percent.

(2) Seamless.

(3) Alloy UNS S30400 or S31600.

(4) Bright annealed.

(5) Passivated per ASTM A380.

KSC-SPEC-Z-0007, Tubing, Steel, Corrosion Resistant, Types 304 and 316, Seamless, Annealed, Specification for, should be used as a guide to standardize procurement.

b. When directly exposed to a marine or launch-induced environment, tubing shall consist of superaustenitic stainless steel (trade name AL6XN) in accordance with ASTM A249, Standard Specification for Welded Austenitic Steel Boiler,

Superheater, Heat Exchanger, and Condenser Tubes, with the following specifications:

(1) Maximum surface roughness of 32 μin on the inside and 63 μin on the outside.

(2) Bright annealed.

(3) Passivated per ASTM A380.

(4) Inspected using Eddy Current Examination per ASTM E426, Standard Practice for Electromagnetic (Eddy-Current) Examination of Seamless and Welded Tubular Products, Austenitic Stainless Steel and Similar Alloys; and Air Underwater Pressure Test per ASTM A1016, Standard Specification for General Requirements for Ferritic Alloy Steel, Austenitic Alloy Steel, and Stainless Steel Tubes.

KSC-SPEC-P-0027, Tubing, Superaustenitic Steel, Corrosion Resistant, UNS N08367 and UNS S31254,Seamed, Bright Annealed, Passivated, Specification for, should be used as a guide to standardize procurement.

5.4.3 Pipe

5.4.3.1 Stainless Steel Pipe

a. Stainless steel pipe for fluid systems shall be in accordance with ASTM A312, Standard Specification for Seamless, Welded, and Heavily Cold Worked Austenitic Stainless Steel Pipes, Types UNS S30400, S30403, S31600, or S31603 (304, 304L, 316, or 316L).

b. Stainless steel pipe for fluid systems with pressures over 1.14 MPa (165 psia), cryogenic systems, hypergolic systems, and other toxic fluid systems shall be seamless.

5.4.3.2 Aluminum Pipe

Aluminum pipe shall be in accordance with ASTM B241, Standard Specification for Aluminum and Aluminum-Alloy Seamless Pipe and Seamless Extruded Tube.

5.4.3.3 Expansion Joints

Expansion joints used in fluid systems in marine- or launch-induced environments shall be made from UNS N06022 (Hastelloy C22) material.

5.4.4 Fluid Fittings

Flared tubing fittings, tube weld fittings, and pipe fittings shall be selected in accordance with KSC-GP-425, Fluid Fitting Engineering Standards, or ASME B16.5, Pipe Flanges and Flanged Fittings NPS 1/2 Through NPS 24 Metric/Inch Standard or ASME B16.9, Factory-Made Wrought Buttwelding Fittings.

Socket weld fittings should not be used for GSE design due to structural and contamination/cleaning concerns.

5.4.5 Fluid System Protective Covers

Protective covers shall be provided for all hoses, ports, fittings, and other fluid-fitting connections to GSE to protect the threads and sealing surfaces and maintain the cleanliness of the system.

Caution should be used in selecting caps and plugs as covers due to the potential for generating debris during installation and removal, especially in oxygen systems. When possible, the protective cover should be connected with a lanyard or the equipment should have a designated storage provision.

5.4.6 Fluid System Components

Fluid system components shall be documented with the following information: Commodity, environment, performance requirements (e.g., pressure rating, flow capacity, etc.), installed dimensions, connection interfaces, recommended vendor, materials, compatibility, qualification/acceptance criteria, and recommended maintenance.

Fluid system components used in the design of liquid or gas systems should be selected from 79K80000, Fluid Component Specifications.

5.4.7 Electrical Power Receptacles and Plugs

Electrical power receptacles and plugs for GSE that interface with other ground systems or facilities shall be in accordance with NFPA 70.

For GSE designed for use at KSC, KSC-STD-E-0011, Electrical Power Receptacles and Plugs, Standard for, should be used as a guide to standardize design practices and ensure compatibility with KSC facility power interfaces.

5.4.8 Electrical Power Cable

Power cables that interface with other ground systems or facilities shall be in accordance with NFPA 70.

5.4.9 Electrical Cable and Harnesses

5.4.9.1 Electrical Cables and Harnesses for Critical Applications

Electrical cables fabricated for critical applications shall be fabricated in accordance with NASA-STD-8739.4, Crimping, Interconnecting Cables, Harnesses, and Wiring, and IPC J-STD-001ES, Space Applications Electronic Hardware Addendum to J-STD-001, Requirements for Soldered Electrical and Electronic Assemblies.

KSC-GP-864, Electrical Ground Support Equipment Cable Handbook, Volume 2A, should be used as a guide to standardize design practices.

5.4.9.2 Electrical Cables and Harnesses for Noncritical Applications

Electrical cables fabricated for noncritical applications shall be fabricated in accordance with IPC-WHMA-A-620, Requirements and Acceptance for Cable and Wire Harness Assemblies, Performance Class 3.

NASA-STD-8739.4 or KSC-GP-864 may be used in lieu of IPC-WHMA-A-620.

5.4.10 Fiber Optics

5.4.10.1 Fiber-Optic Protective Caps

Protective caps shall be provided for all fiber-optic connections to GSE so that the mating surface is protected.

5.4.10.2 Fiber-Optic Cable Assemblies

Fiber-optic cable assemblies, installations, and terminations shall be in accordance with NASA-STD-8739.5, Fiber Optic Terminations, Cable Assemblies, and Installation.

5.4.10.3 Underground Fiber-Optic Cable

Fiber-optic cable for underground cable ducts or direct bury applications shall be in accordance with Telcordia GR-20, Generic Requirements for Optical Fiber and Optical Fiber Cable.

5.4.10.4 Intra-Building Fiber-Optic Cable

Fiber-optic cable for intra-building premise applications shall be in accordance with Telcordia GR-409, Generic Requirements for Indoor Premises Fiber Optic Cable.

5.4.10.5 Single-Mode Fiber-Optic Cable

Single-mode fiber-optic applications shall be in accordance with ITU-T G.652, Characteristics of a Single-Mode Optical Fibre and Cable.

5.4.10.6 Multimode Fiber-Optic Cable

Multimode fiber-optic applications shall be in accordance with ITU-T G.651, Characteristics of a 50/125 Micrometer Multimode Graded Index Optical Fibre Cable.

5.4.11 Electrical Hookup Wire

Electrical hookup wire shall be in accordance with SAE AS 50861, Wire, Electric, Polyvinyl Chloride Insulated, Copper or Copper Alloy; MIL-DTL-16878, Wire, Electrical, Insulated, General Specification for; or SAE AS 22759, Wire, Electrical, Fluoropolymer-Insulated, Copper or Copper Alloy.

5.4.12 Connectors

a. Electrical connectors for GSE used for electrical control and monitoring shall be selected from the following:

(1) SAE AS 50151, Connectors, Electrical, Circular Threaded, AN Type, General Specification for.

(2) MIL-DTL-22992, Connectors, Plugs and Receptacles, Electrical, Waterproof, Quick Disconnect, Heavy Duty Type, General Specification for.

(3) MIL-DTL-24308, Connectors, Electric, Rectangular, Nonenvironmental, Miniature, Polarized Shell, Rack and Panel, General Specification for.

(4) MIL-DTL-38999, Connectors, Electrical, Circular, Miniature, High Density, Quick Disconnect (Bayonet, Threaded and Breech Coupling), Environment Resistant, Removable Crimp and Hermetic Solder Contacts, General Specification for.

(5) IEC 60807, Rectangular Connectors for Frequencies Below 3 MHz.

b. Electrical connectors used in hazardous locations, shall comply with NFPA 70, Article 500.

KSC-GP-864, Volume 2A should be used as a guide to standardize design practices. Scoop-proof electrical connectors should be used in non-hazardous areas. Clocking/ keying electrical connectors should be considered when mismating could cause a hazard. Blind electrical mates should be avoided.

APPROVED FOR PUBLIC RELEASE – DISTRIBUTION IS UNLIMITED

5.4.13 Coaxial Radio Frequency (RF) Connectors

Coaxial RF connectors shall be selected from MIL-PRF-39012, Connectors, Coaxial, Radio Frequency, General Specification for.

5.4.14 Electrical Connector Protective Covers or Dust Caps

Connectors subject to frequent disconnection and connectors that are exposed to harmful environments when disconnected shall have attached caps, plugs, or covers to protect the connectors from damage or contamination while unmated.

Protective covers or caps should have the following characteristics:

- *Protect against moisture intrusion.*

- *Protect sealing surfaces, threads, and pins against damage.*

- *Resist abrasion, chipping, or flaking.*

- *Comply with cleanliness requirements for the plugs and receptacles on which they are used.*

- *Consist of material that is compatible with the connector materials.*

- *Connect to the cable with a suitable lanyard, chain, hinge, or designated storage provision.*

- *Not produce static.*

5.4.15 Optical Covers or Caps

Covers/caps shall be provided to protect optics.

Optical covers/caps should be easily removable for use, as well as easily installable during handling and shipment. The covers/caps should be connected with a lanyard, chain, hinge, or a designated storage provision. For contamination-sensitive surfaces, covers or caps should be maintained clean or cleaned prior to re-installation so that they do not contaminate surfaces being protected.

5.4.16 Sensors and Transducers

Sensors and transducers shall be documented with the following information: Commodity, environment, performance, installed dimensions, connection interfaces, recommended vendor, materials, compatibility, and qualification/acceptance criteria.

Sensors and transducers used in the design of GSE systems should be selected using KSC-NE-9187, Sensors, Transducers and Signal Conditioning Systems Selection Process.

5.4.17 Purged Electrical Enclosures

Purged electrical enclosures shall be in accordance with NFPA 496, Standard for Purged and Pressurized Enclosures for Electrical Equipment.

5.4.18 Racks, Panels, and Modular Enclosures

Electronic racks, panels, and modular enclosures shall conform to the configuration and dimensional requirements of EIA/ECA 310, Cabinets, Racks, Panels, and Associated Equipment.

5.4.19 Printed Circuit Boards (PCBs)

5.4.19.1 PCB Design

Rigid, flexible, and rigid-flex PCBs (single, double, metal-core, or multilayer structures) shall meet the design specifications of the following standards, as applicable:

      a. IPC-2221, Generic Standard on Printed Board Design, Performance Class 3.

      b. IPC-2222, Sectional Design Standard for Rigid Organic Printed Boards.

      c. IPC-2223, Sectional Design Standard for Flexible Printed Boards.

      d. IPC-2252, Design Guide for RF/Microwave Circuit Boards.

5.4.19.2 PCB Fabrication and Acceptance

Rigid, flexible, and rigid-flex PCBs (single, double, metal-core, or multilayer structures) shall meet the qualification and performance specifications of the following standards, as applicable:

      a. IPC-6011, Generic Performance Specification for Printed Boards, Performance Class 3.

      b. IPC-6012, Qualification and Performance Specification for Rigid Printed Boards, Performance Class 3/A (Space and Military Avionics).

c. IPC-6013, Qualification and Performance Specification for Flexible Printed Boards, Performance Class 3.

d. IPC-6018, Microwave End Product Board Inspection and Test, Performance Class 3.

5.4.19.3 Printed Circuit Assembly (PCA) Fabrication and Acceptance

a. PCAs that will not be exposed to the launch-induced environment shall be fabricated in accordance with IPC J-STD-001, Requirements for Soldered Electrical and Electronic Assemblies, Performance Class 3.

b. PCAs that will be exposed to the launch-induced environment shall be fabricated in accordance with NASA-STD-8739.1, Workmanship Standard for Polymeric Application on Electronic Assemblies.

5.4.20 Electric Motors

a. Motors used in GSE shall be in accordance with NEMA MG 1, Motors and Generators.

b. Starters and controllers shall be in accordance with the NEMA standards for industrial control as specified in NEMA ICS 2, Industrial Control and Systems: Controllers, Contactors, and Overload Relays, Rated 600 Volts; Part 8: Disconnect Devices for Use in Industrial Control Equipment; and NFPA 70.

5.5 Electromagnetic Interference (EMI)

Electrical and electronic GSE systems shall be designed to perform when exposed to a minimum level of 20 volts per meter (V/m) in the frequency range from 30 Hz to 18 GHz in accordance with the electromagnetic compatibility requirements in MIL-STD-461, Requirements for the Control of Electromagnetic Interference Characteristics of Subsystems and Equipment.

Program requirements, or the local EMI environment for a specific application, may require a level higher than 20 V/m. The application of MIL-STD-461 to GSE systems should be based on an evaluation of the potential for flight hardware interaction and any existing commercial standards to which the hardware is already certified.

5.6 Identification Markings and Labels

GSE should be painted with the colors in Appendix A by application or fluid commodity for ease of identification and consistency.

5.6.1 Systems and Equipment

a. GSE shall be identified and marked in accordance with MIL-STD-130, Identification Marking of U.S. Military Property.

b. When data matrix identification symbols are used to identify GSE, they shall be applied in accordance with NASA-STD-6002, Applying Data Matrix Identification Symbols on Aerospace Parts.

5.6.2 Load Test Marking

GSE that has been load tested shall be identified and marked with the following information: Drawing/part number and serial number, safe working load, test load, and date of load test.

5.6.3 Compressed Gas Cylinders

Compressed gas cylinders shall be labeled in accordance with 29 CFR 1910.1200, Hazard Communication and Center Requirements.

5.6.4 Handling Equipment

GSE used for hoisting, transportation, or handling shall be marked, in accordance with NASA-STD-8719.9, to indicate the maximum safe working load.

5.6.5 Electrical Cable and Harness Assemblies

Electrical cable and harness assemblies shall be identified at each end of the cable and/or harness and labeled to show the assembly part number, cable or harness reference designation number, and cable or harness end marking, in accordance with NASA-STD-8739.4.

5.6.6 Serial Numbers

Serial numbers or other unique identifiers shall be marked on those end items, assemblies, or components that contain limited-life parts (e.g., valves or regulators containing lot-traceable materials) or that require periodic inspection, checkout, repair, maintenance, servicing, or calibration (e.g., pressure transducers or gages).

5.7 Interchangeability

Hardware assemblies, components, and parts with the same part number shall be physically and functionally interchangeable.

5.8 Safety

a. All GSE shall be designed and fabricated in accordance with NPR 8715.3, NASA General Safety Program Requirements.

GSE must meet the safety requirements for the installation where it will be used, e.g., KNPR 8715.3 for KSC.

b. A hazard analysis shall be conducted as part of the GSE design process to identify, mitigate, and control hazards.

5.9 Human Factors

HF-STD-001, Human Factors Design Standard, shall be used to establish human factors criteria for GSE design.

GSE should include design features to prevent errors or damage during configuration, installation, setup, securing, or safing of a system, such as the following:

- *Dissimilar fittings/connectors to prevent cross-connecting fluid lines or electrical cables.*

- *Conspicuous markings to clearly identify proper mating of parts.*

- *Stenciling of schematics on control panels to assist in setup.*

- *Use of unique identifiers on components and assemblies for ease of identification.*

- *Incorporation of proper warning labels for securing and safing operations.*

- *Protection to prevent inadvertent damage from adjacent operations.*

- *Direct line of sight by the operator to any fittings and connections.*

- *Easy to install, adjustable access platforms, kick plates, etc.*

- *Use of sensors to provide feedback on system configuration to the operator.*

GSE intended for use by operators wearing personnel protective equipment (PPE) should be designed to meet the following criteria:

- *Items (valves, gages, levers, bolts, nuts, and any other items required to be moved, turned, manipulated, or monitored) should be located in a position that will make it easier for a PHE-suited operator to access the item while standing.*

- *Sufficient clearance should be provided to preclude the operator from brushing against other surfaces.*

- *GSE should be designed to avoid requirements for PHE-suited operators to reach into tight areas; stoop to avoid low overhead obstructions; mount supplementary ladders or stairs; touch rough surfaces; or sit, kneel, or lie on the floors or decks.*

- *The design should include suitable provisions to prevent causing discomfort to and fatiguing the PHE-suited personnel.*

- *GSE should be designed to avoid sharp edges, such as expanded metal that could cut or damage the PHE.*

5.10 Security of Information Technology

The design of GSE shall meet the IT security requirements in NPR 2810.1, Security of Information Technology.

6. MATERIALS AND PROCESSES (M&P) REQUIREMENTS

M&P used in interfacing GSE shall be controlled to prevent damage to or contamination of flight hardware.

M&P used in the design and fabrication of GSE should be selected by considering the worst-case operational requirements for the particular application and the design engineering properties of the candidate materials. For example, the operational requirements should include the following: Operational temperature limits, loads, contamination, life expectancy, exposure to moisture or other fluids, and vehicle-related induced and natural environments. Properties that should be considered in material selection include the following: Mechanical properties, fracture toughness, flammability and offgassing characteristics, corrosion, stress corrosion, thermal and mechanical fatigue properties, glass-transition temperature, coefficient of thermal expansion mismatch, vacuum outgassing, fluids compatibility, microbial resistance, moisture resistance, fretting, galling, and susceptibility to electrostatic discharge (ESD) and contamination.

6.1 Material Properties Design Data

a. Material properties used in GSE design shall be determined from test data or a well-documented published source.

The following sources should be used to establish materials properties for materials covered by each document.

> *(1) MMPDS, Metallic Materials Properties Development and Standardization (MMPDS) for metals.*
>
> *(2) MIL-HDBK-17-2, Composite Materials Handbook, Volume 2. Polymer Matrix Composites Materials Properties.*
>
> *(3) MIL-HDBK-17-4, Composite Materials Handbook, Volume 4. Metal Matrix Composites.*
>
> *(4) MIL-HDBK-17-5, Composite Materials Handbook, Volume 5. Ceramic Matrix Composites.*
>
> *(5) Voluntary consensus standard code or standard (e.g., the ASME Boiler and Pressure Vessel Code for pressure vessels and AISC 325 for structural steel).*

The values listed in the codes or standards are minimum material properties. The use of minimum material properties, as stated by the code, is intrinsic to the factor of safety, margin of safety, strength factor, etc., of the design.

b. Parts machined from PCTFE shall comply with ASTM D7194, Standard Specification for Aerospace Parts Machined from Polychlorotrifluoroethylene (PCTFE).

When mechanical properties of new or existing structural materials are not available, they should be determined by the analytical methods described in:

> *(1) MIL-HDBK-17-1, Composite Materials Handbook, Volume 1. Polymer Matrix Composites Guidelines for Characterization of Structural Materials, and MIL-HDBK-17-3, Composite Materials Handbook, Volume 3. Polymer Matrix Composites Materials Usage, Design, and Analysis (for polymers).*
>
> *(2) MIL-HDBK-17-4 for metal matrix composites.*
>
> *(3) MIL-HDBK-17-5 for ceramic matrix composites.*
>
> *(4) MIL-HDBK-149, Rubber (for elastomers).*
>
> *(5) MIL-HDBK-700, Plastics (for polymers).*

If the material is not covered by a design code or one of these sources, the Aerospace Structural Metals Database or other published industry sources should be used in accordance with the best practices for design. The properties listed in these documents are typical values, not minimum values; this must be considered when applying the factor of safety appropriate for the design.

APPROVED FOR PUBLIC RELEASE – DISTRIBUTION IS UNLIMITED

6.2 M&P Controls

M&P controls shall be as follows:

a. All materials and processes shall be defined by standards and specifications and be identified directly on the appropriate engineering drawing.

b. The design drawings shall be approved by an M&P engineer authorized by NASA.

c. Composition and properties of all materials and parts shall be certified by the manufacturer or supplier as required by the procuring document.

d. The Materials and Processes Technical Information System (MAPTIS) shall be consulted to obtain material codes and ratings for materials, standard and commercial parts, and components.

The M&P organization's approval on the engineering drawing approves deviations from the M&P requirements of this Standard. The use of materials and processes that do not comply with the requirements of this Standard may still be acceptable in the actual hardware applications.

6.3 Materials

6.3.1 Flammability and Compatibility Requirements

6.3.1.1 Flammability Control

a. Materials used for flammability control shall be nonflammable or self-extinguishing in their use configuration as defined by NASA-STD-6001, Flammability, Offgassing, and Compatibility Requirements and Test Procedures, Test 1 or Test 10.

Material flammability ratings and tests based on NASA-STD-6001 may be found in the MAPTIS database for many materials.

b. The following materials or methods are also acceptable:

(1) Use of ceramics, metal oxides, and inorganic glasses shall be acceptable without prior testing.

When a material is sufficiently chemically and physically similar to a material found to be acceptable by testing in accordance with NASA-STD-6001, this material may be used without additional testing if its use is approved by the M&P organization.

(2) Materials whose flammability and self-extinguishing properties have been tested in accordance with NASA-STD-6001 under conditions more severe than those encountered in the use environment shall be acceptable without further testing, as in the following examples:

APPROVED FOR PUBLIC RELEASE – DISTRIBUTION IS UNLIMITED

A. Materials used in an environment with an oxygen concentration lower than the test level shall be accepted without testing (provided that the oxygen partial pressure is not substantially greater than the partial pressure at the test level).

B. Materials used in an environment where the oxygen concentration is greater than the test level shall be tested or considered flammable by default.

C. If a material passes the flammability test on a metal substrate, it shall be used on metal substrates of the same thickness or greater.

D. If the material will be used on a thinner or non-heat-sinking substrate (or on no substrate at all), it shall be retested or considered flammable by default.

Materials that are considered flammable by default may still be accepted with approval by the M&P organization.

Many situations arise in which flammable materials are used in an acceptable manner without testing, using mitigation practices and approval by the M&P organization. Guidelines for assessment and mitigation of hardware flammability characteristics can be found in JSC 29353, Flammability Configuration Analysis for Spacecraft Applications.

6.3.1.2 Electrical Wire Insulation Materials

a. Electrical wire insulation materials shall be evaluated for flammability in accordance with NASA-STD-6001, Test 1 or Test 4.

b. Arc tracking shall be evaluated in accordance with NASA-STD-6001, Test 18.

Arc tracking testing is not required for polytetrafluoroethylene (PTFE), PTFE laminate, ethylene tetrafluoroethlyene (ETFE), or silicone-insulated wires since the resistance of these materials to arc tracking has already been established.

6.3.1.3 Fluid Compatibility

6.3.1.3.1 Fluids Other than Oxygen

a. Materials exposed to hazardous fluids other than oxygen (fluids that could cause corrosion, chemically or physically degrade materials in the system, or cause an exothermic reaction) shall be evaluated or tested for compatibility.

NASA-STD-6001, Test 15, is a screening test for short-term exposure to the hydrazine family of fuels, nitrogen tetroxide and mixed oxides of nitrogen and ammonia. For many materials, material compatibility ratings and test results are available in the MAPTIS database.

b. Appropriate compatibility tests shall be conducted for materials that are subjected to long-term exposure to fuels, oxidizers, and other hazardous fluids.

c. The test conditions shall simulate the worst-case use environment that would enhance reactions or degradation of the material or fluid.

d. Materials degradation in long-term tests shall be characterized by post-test analyses of the material and fluid to determine the extent of changes in chemical and physical characteristics, including mechanical properties.

6.3.1.3.2 Oxygen Systems

a. Liquid and gaseous oxygen (LO_2/GO_2) systems shall use materials that are nonflammable in their worst-case use configuration, as defined by NASA-STD-6001, Test 17, for upward flammability in GO_2 (or Test 1 for materials used in oxygen pressures that are less than 350 kPa (50 psia).

Material flammability ratings and test results based on NASA-STD-6001 are found in the MAPTIS database for many materials. KTI-5210, Material Selection List for Oxygen Service, may be referenced for a summary of test results for various materials used in LO_2 and GO_2 applications.

b. When a material in an oxygen system is determined to be flammable by Test 17, an oxygen compatibility assessment shall be conducted in accordance with NASA-STD-6001, and the system safety rationale documented and approved by the M&P organization.

c. When the oxygen compatibility assessment shows the risk is above an acceptable level, configurational testing shall be conducted to support the oxygen compatibility assessment.

d. Configurational testing shall exercise the ignition mechanism in question using an accepted test method.

e. The configurational test method and acceptance criteria shall be reviewed and approved by the M&P organization.

f. The as-built configuration shall be verified against the oxygen compatibility assessment to ensure that mitigation methods identified in the report were incorporated into the design and construction of the hardware.

g. For compressed air systems and pressurized systems containing enriched oxygen, the need for an oxygen compatibility assessment shall be addressed on a case-by-case basis.

Compressed air systems and pressurized systems containing enriched oxygen are inherently less hazardous than systems containing pure oxygen; the hazard increases with oxygen concentration. Pressure hazards exist in all pressurized systems.

APPROVED FOR PUBLIC RELEASE – DISTRIBUTION IS UNLIMITED

Guidelines on the design of safe oxygen systems are contained in ASTM MNL 36, Safe Use of Oxygen and Oxygen Systems: Handbook for Design, Operation, and Maintenance; ASTM G88, Standard Guide for Designing Systems for Oxygen Service; ASTM G63, Standard Guide for Evaluating Nonmetallic Materials for Oxygen Service; ASTM G94, Standard Guide for Evaluating Metals for Oxygen Service; and NASA/TM-2007-213740, Guide for Oxygen Compatibility Assessments on Oxygen Components and Systems.

h. Oxygen and enriched air system components that operate at pressures above 1.83 MPa (265 psia) shall undergo oxygen compatibility acceptance testing at maximum design pressure for a minimum of 10 cycles to ensure that all oxygen system GSE is exposed to oxygen before being connected to flight hardware.

i. Components shall be retested if the results are invalidated by actions occurring after the test (such as rework, repair, or interfacing with hardware for which the cleanliness level is unknown or uncontrolled).

6.3.1.4   Metals

6.3.1.4.1   Steel

Carbon and low alloy steels heat-treated to strength levels at or above 1,240 MPa (180 ksi) tensile strength shall be approved by the M&P organization due to sensitivity to stress corrosion cracking (SCC).

The ductile-to-brittle transition temperature exhibited in steels should be considered when using carbon and low alloy steels in hardware operating in or exposed to low temperatures while in service. For some alloys, the transition temperature may be as high as the ambient temperature.

6.3.1.4.2   Corrosion-Resistant Steel

a. Unstabilized austenitic steels shall not be used under conditions where the temperature is above 371 °C (700 °F).

b. Welding shall be performed only on low carbon, stabilized grades, or superaustenitic grades (e.g., UNS S30403, UNS S31603, UNS S32100, UNS S34700, UNS N08367, and UNS S31254).

Caution should be exercised in using 400 series stainless steels to minimize hydrogen embrittlement, corrosion, and stress corrosion. Austenitic stainless steels are susceptible to pitting corrosion and crevice corrosion in a chloride-rich (marine) environment; some austenitic stainless steels are susceptible to SCC in a chloride-rich (marine) environment.

c. Free-machining alloys such as UNS S30300 and UNS S30323 shall not be used in natural or launch-induced environments due to service-related corrosion issues.

d. UNS N08367 or UNS S31254 shall be used for tubing exposed to marine and launch-induced environments.

e. Cleaning, descaling, and passivating of stainless steel parts, assemblies, equipment, and installed systems shall be in accordance with ASTM A380, Standard Practice for Cleaning, Descaling, and Passivation of Stainless Steel Parts, Equipment, and Systems.

f. When acid cleaning baths are used for steel parts, the parts shall be baked in accordance with SAE AMS 2759/9, Hydrogen Embrittlement Relief (Baking) of Steel Parts, to alleviate potential hydrogen embrittlement problems.

Hardware should be designed to avoid fretting and/or wear of stainless steel alloys. Lubricants and lubricated coatings should be considered for use with stainless steel materials in applications where they come into contact with each other through a sliding movement, and gall-resistant alloys such as Nitronic® should be considered as alternatives.

6.3.1.4.3 Aluminum

a. Aluminum alloys used in structural applications shall be resistant to general corrosion, pitting, intergranular corrosion, and SCC.

b. 5000-series alloys containing more than 3 percent magnesium shall not be used in applications where the temperature exceeds 66 °C (150 °F), because grain boundary precipitation above this temperature can create stress-corrosion sensitivity.

Hardware made with aluminum alloys should not be loaded through the short transverse grain direction, as resistance to SCC is at a minimum in that direction.

6.3.1.4.4 Nickel-Based Alloys

Alloys with a high nickel content are susceptible to sulfur embrittlement; therefore, any foreign material that could contain sulfur, such as oils, grease, and cutting lubricants, shall be removed prior to heat treatment, welding, or high temperature service.

Some of the precipitation-hardening superalloys are susceptible to depletion of the alloying element at the surface in a high temperature, oxidizing environment. This effect should be carefully evaluated when a thin sheet is used, since a slight depletion could involve a considerable proportion of the cross section of the material.

6.3.1.4.5 Titanium

a. Areas subject to fretting and/or wear shall be anodized in accordance with SAE AMS 2488, Anodic Treatment–Titanium and Titanium Alloys Solution pH 13 or Higher, or hard-coated using a wear-resistant material such as a tungsten carbide/cobalt thermal spray.

Titanium and its alloys exhibit very poor resistance to wear. Fretting that occurs at interfaces with titanium and its alloys often cause cracks to occur, especially due to fatigue. The preferred policy is to implement a design that precludes the fretting and/or wear that occurs with titanium and its alloys.

b. Titanium alloys shall not be used with LO_2 or GO_2 at any pressure or with air at oxygen partial pressures above 35 kPa (5 psia).

c. The surfaces of titanium and titanium alloy mill products shall be 100-percent machined, chemically milled, or pickled to a sufficient depth to remove all contaminated zones and layers formed while the material was exposed to elevated temperatures.

Contaminated zones and layers may be formed as a result of mill processing, heat treating, and forming operations at elevated temperatures.

d. All cleaning fluids and other chemicals used during manufacturing and processing of titanium hardware shall be verified to be compatible before use and removed prior to heat treatment.

The use of titanium in hydrochloric acid, chlorinated solvents, chlorinated cutting fluids, fluorinated hydrocarbons, and anhydrous methyl alcohol should be avoided due to titanium's susceptibility to SCC. Contact of titanium alloys with mercury, cadmium, silver, and gold should be avoided at certain temperature ranges because of liquid-metal-induced embrittlement and/or solid-metal-induced embrittlement. Hardware should be designed to avoid fretting and/or wear of titanium alloys.

6.3.1.4.6 Copper Alloys

a. Copper alloys, such as brasses and bronzes, shall be resistant to corrosion, pitting, and SCC.

b. To prevent SCC, copper-based alloys such as brass shall not be used in solutions with ammonium ions or in contact with ammonia.

GSE should be designed so that copper is not exposed to hydrazine environments. Copper has the potential for SCC when exposed to ammonia, which is a product of hydrazine decomposition.

Beryllium copper (UNS C17200) is commonly used for high-strength, nonsparking structural components in applications where it is subject to contact and wear.

6.3.1.4.7 Beryllium and Beryllium Alloys

Beryllium particles, beryllium oxide, and other beryllium compounds are toxic when inhaled. Extreme caution must be exercised during fabrication to avoid exposing personnel to beryllium or beryllium compounds.

APPROVED FOR PUBLIC RELEASE – DISTRIBUTION IS UNLIMITED

Machining, grinding, and finishing operations on beryllium and beryllium alloys shall be performed either wet, using a liquid coolant with local ventilation, or dry, using high-velocity, close-capture ventilation.

Consult the appropriate Material Safety Data Sheet and health, safety, and environmental documentation for more information.

6.3.1.4.8 Tin

a. Tin and tin plating shall not be used in an application unless the tin is alloyed with at least 3 percent lead to prevent the growth of tin whiskers.

b. For critical GSE, lot sampling shall be used to verify the presence of at least 3 percent lead.

6.3.1.5 Nonmetals

6.3.1.5.1 Elastomers

a. Elastomeric materials shall be selected to operate within the parameters of a design service life, including the vendor-specified shelf life.

b. Elastomeric materials shall be cure-dated for tracking purposes.

c. Elastomers shall not have a corrosive effect on other materials when exposed to conditions normally encountered in service.

Examples include one-part silicones that liberate acetic acid when they are cured.

d. When rubbers or elastomers are used at low temperatures, the ability of these materials to maintain the required elastomeric properties shall be verified by testing them at or below use temperature.

Elastomers used in GSE should be in accordance with MIL-HDBK-149.

6.3.1.5.2 Composite Materials

Defects resulting from the manufacturing process shall be assessed through nondestructive evaluation (NDE) techniques to meet the intent of paragraph 6.4.13 of this Standard.

Composite materials used in GSE should be developed and qualified in accordance with MIL-HDBK-17, Volumes 1 through 5.

6.3.1.5.3 Lubricants

NASA-TM-86556, Lubrication Handbook for the Space Industry, Part A: Solid Lubricants and Part B, Liquid Lubricants, should be used in evaluating and selecting lubricants for GSE. Lubricants are not restricted to those listed in NASA-TM-86556; guidelines on additional lubricants are contained in NASA/CR-2005-213424, Lubrication for Space Applications. Long-life performance should be considered when selecting lubricants. Use of lubricants in close proximity to precision-cleaned hardware or electrical connections should be minimized or tightly controlled to prevent cross-contamination.

a. Lubricants containing chloro-fluoro components shall not be used with aluminum or magnesium in applications where fretting, vibration, or wear can occur.

b. Hardware with lubricants containing chloro-fluoro components shall not be heated above the maximum rated temperature for the lubricant.

Decomposition/reaction products from over-heating lubricants with chloro-fluoro components can attack metallic materials and can be toxic to personnel.

c. Lubrication of flared tube fittings shall be in accordance with SAE AIR 4071, Lubricants for Oxygen Use.

6.3.1.5.4 Limited-Life Materials

Materials that do not meet the design life requirements shall be identified as limited-life items requiring maintenance or replacement.

6.3.1.5.5 Plastic Films, Foams, and Adhesive Tapes

Thin plastic films and tape materials used in GSE shall be tested in accordance with and meet the requirements of the following:

a. Flammability testing per NASA-STD-6001 if the GSE could be exposed to ignition sources.

b. Electrostatic discharge testing per MMA-1985-79, Standard Test Method for Evaluating Triboelectric Charge Generation and Decay, if the GSE could be used or stored near ESD-sensitive materials or equipment.

c. Reactivity testing per NASA-STD-6001, A.7 Reactivity and Penetration of Materials due to Incidental Exposure to Hydrazine, Monomethylhydrazine, Unsymmetrical Dimethylhydrazine, Aerozine 50, Nitrogen Tetroxide, and Ammonia, for hypergolic ignition/breakthrough characteristics if the GSE could be exposed to one of the listed fluids.

Material flammability ratings for many Plastic Films, Foams, and Adhesive Tapes are found in the MAPTIS database. KTI-5212, Material Selection List for Plastic Films, Foams, and Adhesive Tapes, may be referenced for a summary of flammability, ESD, and hypergol compatibility test results for various materials.

6.3.1.5.6 Fungus Resistance

For GSE used in a marine environment, materials that do not provide nutrients to fungi shall be used.

Materials that do not provide nutrients to fungi should be selected from MIL-HDBK-454, General Guidelines for Electronic Equipment, Table 4-I, Group I, except when one of the following criteria is met:

a. Materials are used inside environmentally sealed containers with an internal container humidity of less than 60 percent relative humidity (RH) at ambient conditions.

b. Materials are used inside electrical boxes where the temperature is always greater than or equal to the ambient cabin temperature.

c. Only the edges of materials are exposed.

d. Materials are fluorocarbon polymers (including ETFE) or silicones.

Alternate materials should be tested for fungus resistance in accordance with MIL-STD-810. When materials that do provide nutrients to fungi are used, they should be treated to prevent fungus growth. Materials not treated for fungus growth should be identified on the drawing, as well as any action required, such as periodic inspection, maintenance, or replacement of the material. Treatment for fungus growth should not adversely affect performance or service life or constitute a health hazard. Materials treated for fungus resistance should be protected from environments that would leach out the protective agent.

6.4 Processes

Riveting should be in accordance with MSFC-STD-156, Riveting, Fabrication and Inspection, Standard for.

6.4.1 Welding

Fusion welding of GSE shall meet all requirements for nonflight hardware in AWS D17.1, Specification for Fusion Welding for Aerospace Applications, with the exception of pressure systems.

Welding of pressure systems is covered by ASME BPVC and ASME B31.3. The selection of parent materials and weld methods for GSE should be based on consideration of the weldments, including adjacent heat-affected zones, as they affect the operational capability of the parts concerned. Welding procedures should be selected to provide a weld of the required quality, use the minimum amount of energy, and protect the heated metal from contamination.

6.4.1.1   Welding Procedure Qualification

a. Prior to the start of production welding, a welding procedure specification (WPS) shall be prepared and qualified for each weld to be made.

A WPS is always required, even if the procedure is considered to be prequalified by an applicable AWS standard. WPS's for stainless steels and nickel alloys must always be qualified by testing.

b. Prior to the start of any production welding, all WPS's and procedure qualification records (PQR's) shall be approved by an M&P engineer authorized by NASA.

6.4.1.2   Welder/Welding Operator Performance Qualification

a. Prior to the start of production welding, each welder and welding operator shall be qualified.

b. Prior to the start of any production welding, all welder/welding operator performance qualification (WPQ) shall be approved by an M&P engineer authorized by NASA.

6.4.1.3   Weld Classification and Inspection

a. Welds whose failure could cause personnel injury, damage to flight hardware, loss of mission, or major damage to a significant ground asset shall be classified and inspected as Class A welds.

b. Welds whose failure could cause loss of function or improper performance shall be classified and inspected as Class B welds.

c. Welds whose failure could not cause improper performance or a hazard to personnel shall be classified and inspected as Class C welds.

6.4.2   Brazing

Brazing should be conducted in accordance with AWS C3.3, Recommended Practices for Design, Manufacture, and Examination of Critical Brazed Components.

a. Brazing of aluminum alloys shall meet the requirements of AWS C3.7M/C3.7, Specification for Aluminum Brazing.

b. Torch, induction, and furnace brazing shall meet the requirements of AWS C3.4, Specification for Torch Brazing; AWS C3.5, Specification for Induction Brazing; and AWS C3.6, Specification for Furnace Brazing, respectively.

c. Subsequent fusion welding operations in the vicinity of brazed joints or other operations involving high temperatures that might affect the brazed joint shall be prohibited unless it can be demonstrated that the fixturing, processes, methods, and/or procedures employed will preclude degradation of the brazed joint.

d. Brazed joints shall be designed for shear loading and, for structural parts, not be relied upon for strength in axial loading.

e. The shear strength of brazed joints shall be evaluated in accordance with AWS C3.2M/C3.2, Standard Method for Evaluating the Strength of Brazed Joints.

f. For furnace brazing of complex configurations, such as heat exchangers and cold plates, destructive testing shall be conducted on preproduction brazed joints to verify that the brazed layer that extends beyond the fillet area is continuous and forms a uniform phase.

g. Brazing of pressure vessels shall be in accordance with ASME Boiler and Pressure Vessel Code, Section VIII.

6.4.3 Soldering

a. Soldering of electrical connections not exposed to vibration or thermal cycling shall be performed in accordance with IPC J-STD-001, Requirements for Soldered Electrical and Electronic Assemblies.

b. Soldering of electrical connections exposed to vibration or thermal cycling shall be performed in accordance with IPC J-STD-001ES.

c. All solderable platings and protective finishes based on tin shall contain a minimum lead content of 3 percent by weight.

   (1) For critical GSE, lot sampling shall be used to verify the presence of at least 3 percent lead.

d. Soldering shall not be used for structural applications.

6.4.4 Heat Treating and Plating

a. Heat treatment of aluminum alloy parts shall meet the requirements of SAE AMS 2772, Heat Treatment of Aluminum Alloy Raw Materials; SAE AMS 2770, Heat Treatment of Wrought Aluminum Alloy Parts; or SAE AMS 2771, Heat Treatment of Aluminum Alloy Castings, as appropriate.

b. Steel parts shall be heat-treated to meet the requirements of SAE AMS-H-6875, Heat Treatment of Steel Raw Materials, or SAE AMS 2759, Heat Treatment of Steel Parts General Requirements.

c. Heat treatment of titanium and titanium alloy parts shall meet the requirements of SAE AMS-H-81200, Heat Treatment of Titanium and Titanium Alloys.

d. Heat treatment of nickel-based and cobalt-based alloy parts shall meet the requirements of SAE AMS 2774, Heat Treatment Wrought Nickel Alloy and Cobalt Alloy Parts.

e. Electrodeposited nickel plating shall be applied according to the requirements of SAE AMS 2403, Plating, Nickel, General Purpose, or SAE AMS 2423, Plating, Nickel, Hard Deposit.

f. Electroless nickel plate shall be applied in accordance with SAE AMS 2404, Plating, Electroless Nickel.

g. The nickel-aluminum interface in nickel-plated aluminum shall be protected from exposure to corrosive environments.

Nickel and aluminum form a strong galvanic cell at the nickel-aluminum interface, and exposure of the aluminum alloy to a corrosive environment can produce rapid debonding of the nickel plate.

h. Galvanized (zinc) coatings shall be applied in accordance with the following:

(1) ASTM A123, Standard Specification for Zinc (Hot-Dip Galvanized) Coatings on Iron and Steel Products, for structural components.

(2) ASTM A153, Standard Specification for Zinc Coating (Hot-Dip) on Iron and Steel Hardware, for associated iron and steel hardware such as nuts, bolts, and washers that are coated by immersing them in molten zinc.

(3) ASTM A653, Standard Specification for Steel Sheet, Zinc-Coated (Galvanized) or Zinc-Iron Alloy-Coated (Galvannealed) by the Hot-Dip Process, for sheet materials.

i.   All repairs to damaged galvanized coatings shall be in accordance with ASTM A780, Standard Practice for Repair of Damaged and Uncoated Areas of Hot-Dip Galvanized Coatings.

6.4.5   Forging

Because mechanical properties are optimum in the direction of material flow during forging, forging techniques should be used that produce an internal grain-flow pattern such that the direction of flow is parallel to the principal stresses. The forging pattern should be free from reentrant and sharply folded flow lines.

For fracture-critical forgings, the following shall apply:

a.   After the forging technique (including degree of working) is established, the first production forging shall be sectioned to show the grain-flow patterns and to verify mechanical properties.

b.   The procedure shall be repeated after any change in the forging technique.

The information gained from this effort should be used to redesign the forging technique as necessary.

c.   The resulting data shall be retained and made available for review by the procuring activity.

6.4.6   Casting

a.   Castings shall meet the requirements in SAE AMS 2175, Castings, Classification and Inspection of.

b.   Fracture-critical castings shall meet the following:

(1) Pre-production castings shall be subjected to first article inspection to verify proper material flow, proper material integrity, minimum required mechanical properties, proper grain size, and macro/microstructure.

(2) The same casting practice and heat-treating procedure shall be used for the production castings as for the approved first-article castings.

(3) Mechanical property testing of integrally cast or excised tensile bars at critical locations shall be conducted to ensure foundry control of cast lots.

(4) Periodic cut-ups or functional testing shall be conducted.

Additional requirements in NASA-STD-5009, Nondestructive Evaluation Requirements for Fracture-Critical Metallic Components, apply to fracture-critical castings (see 6.4.13b).

6.4.7 Adhesive Bonding

Retesting of adhesives used for production parts is not required if they are within the manufacturer's recommended shelf life.

    a.   Structural adhesive bonding processes shall be controlled by a documented process.

The sensitivity of structural adhesive bonds to contamination is of particular concern. In the absence of relevant performance data, bond sensitivity studies should be conducted to verify that the required adhesive properties are maintained after exposure to the expected materials at the expected concentrations, including ozone, ambient humidity, cleaning fluids, and lubricants. Adequate in-process cleanliness inspections should be conducted as part of the bonding process.

    b.   Bonded primary structural joints shall demonstrate cohesive failure modes in shear at ambient temperature.

    c.   Adhesives shall not have a corrosive effect on other materials during cure or when exposed to conditions normally encountered in service.

Structural adhesive bonding should be in accordance with MSFC-SPEC-445, Adhesive Bonding, Process and Inspection, Requirements for, with the exception of retesting.

6.4.8 Fluid System Cleanliness

    a.   Fluid components shall be precision-cleaned to level 300A as a minimum in accordance with ISO 14952, Space Systems Surface Cleanliness of Fluid Systems.

Cleaning to level 300A per KSC-C-123, Surface Cleanliness of Ground Support Equipment Fluid Systems, Specification for, is considered equivalent to ISO 14952.

    b.   For GSE interfaces with precision-cleaned flight fluid systems, supply interface/final filters shall be located as close to the flight hardware interface as possible.

    c.   Interface filters shall be used on outlet lines if it is determined that any operations, such as servicing or deservicing fluids, could permit flow in a reverse direction.

    d.   Interfacing fluid system GSE shall be cleaned to meet or exceed the cleanliness level of the flight hardware.

    e.   Fluid system GSE shall be designed to allow comprehensive cleaning of all interior surfaces.

Socket welds, dead heads, contamination traps, and corrugated surfaces should be avoided. For complex piping systems, the designer should allow for removal of components (e.g. regulator, valves, and gauges) during cleaning. The designer should make provisions for connecting the cleaning equipment to the system and allow for purging and drying. "Flow through" cleaning does not thoroughly clean all wetted surfaces and may result in trapped cleaning fluid that will act as a contaminant. For hypergol systems, residual cleaning fluid may be reactive with the hypergol propellant.

6.4.9 Crimping

Crimping shall be in accordance with NASA-STD-8739.4.

6.4.10 Potting and Molding of Cable Assemblies

Potting and molding of electrical connectors shall be accomplished to provide physical strength, strain relief, and seal to the connector to cable interfaces for cable assemblies.

Potting provides insulated sealing of the electrical pins or sockets inside the rear connector housing. Molding provides physical strength, strain relief, and seal to the cable sheath and the outside of the connector housing. KSC-STD-132, Potting and Molding Electrical Cable Assembly Terminations, Standard for, should be used as a guide to standardize design practices.

6.4.11 Corrosion Control

Protective coating of hardware should be appropriate to the condition, use, and environment to which the hardware will be exposed during its life cycle. The coating should minimize corrosion, and its color indicates its use (see paragraph 4.2.2.3). Guidelines for corrosion control for facilities, systems, and equipment are given in TM-584, Corrosion Control and Treatment Manual.

   a.   Protective coating of hardware shall be in accordance with NASA-STD-5008, Protective Coating of Carbon Steel, Stainless Steel, and Aluminum on Launch Structures, Facilities, and Ground Support Equipment.

   b.   Corrosion control of galvanic couples shall be in accordance with MIL-STD-889, Dissimilar Metals.

   c.   All contacts between graphite-based composites and metallic materials shall be treated as dissimilar metal couples and sealed in accordance with NASA-STD-5008.

   d.   For critical GSE, the following additional requirements shall also be implemented:

   (1) Faying surfaces of metal alloys shall be sealed in accordance with NASA-STD-5008.

(2) The faying surfaces of all electrical bonding connections shall be sealed, except for nickel-plated surfaces.

6.4.12 Stress Corrosion Cracking

Materials exposed to a marine or launch-induced environment shall be selected from alloys that are highly resistant to SCC as specified in MSFC-STD-3029, Guidelines for the Selection of Metallic Materials for Stress Corrosion Cracking Resistance in Sodium Chloride Environments.

6.4.13 Material NDE

NDE of pressure vessels is in accordance with ASME BPCV, and NDE of fluid systems is in accordance with ASME B31.3.

    a. NDE for all components except fracture-critical metallic GSE components shall be performed in accordance with MIL-HDBK-6870, Inspection Program Requirements, Nondestructive for Aircraft and Missile Materials and Parts.

    b. NDE for all fracture-critical metallic GSE components shall be performed in accordance with NASA-STD-5009.

    c. Qualification and certification of personnel involved in NDE shall comply with AIA/NAS 410, NAS Certification & Qualification of Nondestructive Test Personnel.

Typical NDE methods include penetrant, magnetic particle, radiographic, ultrasonic, and eddy current testing. Other NDE inspection methods include leak testing, shearography, and thermography. NDE of welds is covered in section 6.4.1 of this Standard.

6.4.14 Chemical Etching

    a. Metals with surfaces that have been smeared/flowed by processing shall require chemical etching before penetrant inspection.

Processes causing smearing include, but are not limited to, machining, grinding, grit blasting, wire brushing, peening, and polishing.

    b. High-strength steels etched to remove smeared metal shall be baked after etching to prevent hydrogen embrittlement in accordance with SAE AMS-2759/9, Hydrogen Embrittlement Relief (Baking) of Steel Parts.

    c. Threads and holes shall be masked or plugged before etching.

6.4.15 Hydrogen Embrittlement

Hydrogen embrittlement of metals is not very well understood, and only a limited amount of materials property data has been generated and reported in MAPTIS.

    a. Metallic materials used in hydrogen systems in critical GSE shall be approved by the M&P organization.

Test data may have to be generated in a simulated environment to support the rationale. Guidelines for designing safe hydrogen systems are contained in AIAA G-095, Guide to Safety of Hydrogen and Hydrogen Systems.

    b. Metallic materials that are electrochemically treated or exposed to acids or bases during manufacturing or processing shall be processed in a manner to prevent hydrogen embrittlement, or be treated for hydrogen embrittlement relief in accordance with SAE AMS-2759/9.

6.4.16 Contamination Control

    a. A contamination control plan shall be generated in accordance with ASTM E1548, Standard Practice for Preparation of Aerospace Contamination Control Plans.

The contamination control plan should include controls on contamination-sensitive manufacturing processes, such as adhesive bonding; controls on packaging for shipment and storage; cleanliness level acceptance limits and verification methods for fluid systems; and a foreign-object-debris (FOD) prevention program. The FOD prevention program should be established for all mechanical and electrical GSE, including the design, development, manufacturing, assembly, repair, processing, testing, maintenance, operation, and checkout of the equipment to ensure the highest practical level of cleanliness. The FOD prevention program should follow the National Aerospace FOD Prevention Inc. (NAFPI) Guideline, FOD Prevention Guideline. The FOD prevention program should also conform to AIA/NAS 412, Foreign Object Damage/Foreign Object Debris (FOD) Prevention.

    b. Definitions shall be established for cleanliness-level acceptance limits and verification methods for GSE fluid systems and for GSE internal and external surfaces that interface with flight hardware.

    c. Cleanliness requirements for GSE shall be identified in the engineering documentation.

APPENDIX A

GSE IDENTIFICATION COLORS

A.1   Purpose

The purpose of this appendix is to provide the identification colors for GSE, which are presented in the chart below.

Application	Color	FED-STD-595 Code
Electrical/electronic, hydro/pneumatic consoles, racks, and cabinets	Gray	26440 or 26251
Structural steel/aluminum	Gray	16187 or 16473
Remove-before-flight items, safety equipment, and protective equipment	Red	11105 or 21105
White room or clean room equipment and pressure vessels	White	17875 or 27875
Panel lettering	Black	37038
Handling and transportation equipment	Yellow or White	13538, 17875, or 27875
Equipment for hypergolic fuel servicing	Yellow w/Brown Band	13655 (yellow), 10080 (brown)
Equipment for hypergolic oxidizer servicing	Green w/Brown Band	14110 (green), 10080 (brown)
Control racks and consoles	Blue	25102

APPENDIX B

DELIVERABLES

B.1   Purpose

The purpose of this appendix is to provide information on GSE Acceptance Data Package contents.

B.1.1   The GSE provider has the responsibility to submit documentation to verify that the hardware/software has been developed in accordance with this Standard.

B.1.2   The GSE provider has the responsibility to provide all the necessary documentation to the using organization when the GSE is delivered for use, regardless of who "owns" the GSE at the time of delivery.

Examples of this documentation include, but are not limited to, the following:

 a.   Certification Records (indicates how the GSE was certified as complying with this Standard).

 b.   Master Verification Matrix (indicates which GSE requirements were met and how).

 c.   Material Inspection and Receiving Report.

 d.   Validation and verification compliance records.

 e.   Drawings with parts list or bills of material.

 f.   Maintenance manuals/procedures.

 g.   Material certifications and lot traceability.

 h.   Operating manuals/procedures.

 i.   Software/Firmware Version Description document.

 j.   Facility and Flight Vehicle Interface requirements.

 k.   Hazard Analyses or Ground Safety Data pack.

 l.   Failure Modes, Effects, and Criticality Analysis.

 m.   Critical Items List.

Intent/Rationale: The using organization requires documentation for safely operating, maintaining, and servicing the GSE. To reduce risk to the mission, as well as to ground personnel and flight crews, a failure mode and effects analysis should be completed and submitted in accordance with the criticality assigned to the GSE by the responsible program or project.

APPENDIX C

REFERENCE DOCUMENTS AND DATABASE

C.1   Purpose

The purpose of this appendix is to provide guidance and is made available in the reference documents listed below.

C.2   Reference Documents

	SMI - Handbook for Spring Design
79K80000	Fluid Component Specifications
AFSPCMAN 91-710, Volume 3	Range Safety User Requirements Manual Volume 3 - Launch Vehicles, Payloads, and Ground Support Systems Requirements
AIA/NAS 412	Foreign Object Damage/Foreign Object Debris (FOD) Prevention
AIAA R-100	Recommended Practice for Parts Management
AIAA/G-095	Guide to Safety of Hydrogen and Hydrogen Systems
ANSI/ESD S20.20	ESD Association Standard for the Development of an Electrostatic Discharge Control Program for – Protection of Electrical and Electronic Parts, Assemblies and Equipment (Excluding Electrically Initiated Explosive Devices)
API Standard 620	Design and Construction of Large, Welded, Low-pressure Storage Tanks
ASTM A325	Standard Specification for Structural Bolts, Steel, Heat Treated, 120/105 ksi Minimum Tensile Strength
ASTM A490	Standard Specification for Structural Bolts, Alloy Steel, Heat Treated, 150 ksi Minimum Tensile Strength

ASTM D4169	Performance Testing of Shipping Containers and Systems
ASTM D5363	Standard Specification for Anaerobic Single-Component Adhesives (AN)
ASTM G63	Standard Guide for Evaluating Nonmetallic Materials for Oxygen Service
ASTM G88	Standard Guide for Designing Systems for Oxygen Service
ASTM G94	Standard Guide for Evaluating Metals for Oxygen Service
ASTM MNL 36	Safe Use of Oxygen and Oxygen Systems: Handbook for Design, Operation, and Maintenance
AWS C3.3	Recommended Practices for Design, Manufacture, and Examination of Critical Brazed Components
FED-STD-595	Colors Used in Government Procurement
IEEE 446	IEEE Recommended Practice for Emergency and Standby Power Systems for Industrial and Commercial Applications
IEEE 484	IEEE Recommended Practice for Installation Design and Installation of Vented Lead-Acid Batteries for Stationary Applications
IEEE 1106	IEEE Recommended Practice for Installation, Maintenance, Testing, and Replacement of Vented Nickel-Cadmium Batteries for Stationary Applications
IEEE 1187	IEEE Recommended Practice for Installation Design and Installation of Valve-Regulated Lead-Acid Storage Batteries for Stationary Applications
JSC 29353	Flammability Configuration Analysis for Spacecraft Applications
KSC-C-123	Surface Cleanliness of Ground Support Equipment Fluid Systems, Specification for
KSC-DE-512-SM	Facility, System, and Equipment General Design Requirements

KSC-E-165	Electrical Ground Support Equipment Fabrication, Specification for
KSC-GP-864	Volume IIA, Electrical Ground Support Equipment Cable Handbook
KSC-GP-986	KSC Design Criteria for Reusable Space Vehicle Umbilical Systems
KSC-NE-9187	Sensors, Transducers and Signal Conditioning Systems Selection Process
KSC-SPEC-P-0012	Refractory Concrete, Specification for
KSC-SPEC-P-0027	Tubing, Superaustenitic Steel, Corrosion Resistant, UNS N08367 and UNS S31254, Seamed, Bright Annealed, Passivated, Specification for
KSC-SPEC-Z-0007	Tubing, Steel, Corrosion Resistant, Types 304 and 316, Seamless, Annealed, Specification for
KSC-SPEC-Z-0008	Fabrication and Installation of Flared Tube Assemblies and Installation of Fittings and Fitting Assemblies, Specification for
KSC-STD-132	Potting and Molding Electrical Cable Assembly Terminations, Standard for
KSC-STD-164	Environmental Test Methods for Ground Support Equipment, Standard for
KSC-STD-E-0001	Design of Electrical Control and Monitoring Systems, Equipment (GSE) and Panels, Standard for
KSC-STD-E-0002	Hazard Proofing of Electrically Energized Equipment, Standard for
KSC-STD-E-0011	Electrical Power Receptacles and Plugs, Standard for
KSC-STD-E-0012	Facility Grounding and Lightning Protection, Standard for
KSC-STD-SF-0004	Safety Standard for Ground Piping Systems Color Coding and Identification
KSC-STD-Z-0005	Design of Pneumatic Ground-Support Equipment, Standard for

KSC-STD-Z-0006	Hypergolic Propellants Ground Support Equipment, Design of, Standard for
KSC-STD-Z-0008	Design of Ground Life Support Systems and Equipment, Standard for
KSC-STD-Z-0009	Design of Cryogenic Ground Support Equipment, Standard for
KTI-5210	Material Selection List for Oxygen Service
KTI-5212	Material Selection List for Plastic Films, Foams, and Adhesive Tapes
MIL-HDBK-17-1	Composite Materials Handbook Volume 1. Polymer Matrix Composites Guidelines for Characterization of Structural Materials
MIL-HDBK-17-2	Composite Materials Handbook Volume 2. Polymer Matrix Composites Materials Properties
MIL-HDBK-17-3	Composite Materials Handbook Volume 3. Polymer Matrix Composites Materials Usage, Design, and Analysis
MIL-HDBK-17-4	Composite Materials Handbook Volume 4. Metal Matrix Composites
MIL-HDBK-17-5	Composite Materials Handbook Volume 5. Ceramic Matrix Composites
MIL-HDBK-149	Rubber
MIL-HDBK-454	General Guidelines for Electronic Equipment
MIL-HDBK-700	Plastic
MMPDS	Metallic Materials Properties Development and Standardization
MSFC-SPEC-445	Adhesive Bonding, Process and Inspection, Requirements for
MSFC-STD-156	Riveting, Fabrication and Inspection, Standard for

MSFC-STD-486	Standard, Threaded Fasteners, Torque Limits for
NASA/CR-2005-213424	Lubrication for Space Applications
NASA RP-1228	Fastener Design Manual
NASA/TM-2007-213740	Guide for Oxygen Compatibility Assessments on Oxygen Components and Systems
NASA-TM-86556	Lubrication Handbook for the Space Industry, Part A: Solid Lubricants, Part B: Liquid Lubricants
NASA/TP-2003-212242	EEE-INST-002: Instructions for EEE Parts Selection, Screening, Qualification, and Derating
NPD 8730.2	Parts Policy
SSP 50004	Ground Support Equipment Design Requirements: International Space Station
SW-E-0002	Ground Support Equipment General Design Requirements: Space Shuttle
TM-584	Corrosion Control and Treatment Manual

C.3 Reference Database

CINDAS Aerospace Structural Metals Database